REPUBLIC OF SOUTH AFRICA
REPUBLIEK VAN SUID-AFRIKA

DEPARTMENT OF AGRICULTURAL TECHNICAL SERVICES
DEPARTEMENT VAN LANDBOU-TEGNIESE DIENSTE

FLORA OF SOUTHERN AFRICA

VOL. 10, PART 1

ISBN 0 621 04726 0

FLORA OF SOUTHERN AFRICA

which deals with the territories of

SOUTH AFRICA, TRANSKEI, LESOTHO, SWAZILAND, BOPHUTHATSWANA, SOUTH WEST AFRICA/NAMIBIA AND BOTSWANA

VOLUME 10, PART 1

Edited by

O. A. Leistner

Editorial Committee: B. de Winter, D. J. B. Killick and
O. A. Leistner

Botanical Research Institute,
Department of Agricultural Technical Services

1979

CONTENTS

INTRODUCTION

For a key to the families, the Flora should be used in conjunction with Dyer's Genera of Southern African Flowering Plants, Vol. 1 (1975) and Vol. 2 (1976), which are arranged on the lines of the Engler system. The genera are numbered, as far as possible, according to the list published by De Dalla Torre and Harms in their Genera Siphonogamarum (1900–1907) in order to facilitate reference, though genera in the Flora are not necessarily arranged in this sequence.

The following condensed abbreviations for literature references are used:

C.F.A. .... Conspectus Florae Angolensis

Dyer, Gen. .... The Genera of Southern African Flowering Plants by R. A. Dyer, Vol. 1 (1975) and Vol. 2 (1976)

F.C. .... Flora Capensis

F.C.B. .... Flore du Congo et du Rwanda-Burundi

F.S.W.A. .... Prodromus einer Flora von Südwestafrika

F.T.A. .... Flora of Tropical Africa

F.T.E.A. .... Flora of Tropical East Africa

F.W.T.A. .... Flora of West Tropical Africa

F.Z. .... Flora Zambesiaca

Burtt Davy, Fl. Transv. ... Manual of the Flowering Plants and Ferns of the Transvaal and Swaziland, Vol. 1 (1926) and Vol. 2 (1932).

Cited voucher specimens given without indication of herbarium are housed in PRE (National Herbarium, Pretoria).

This part was compiled in accordance with a Guide to Contributors to the Flora of Southern Africa (Ross, Leistner & De Winter, 1977), which is available from the Librarian, Botanical Research Institute, Private Bag X101, Pretoria, 0001.

Volume 10 of the Flora, of which the present publication is a component, will appear in parts (see p. vi). The number of the part, which in the present publication is "1", precedes the page number on all pages marked with Arabic numerals. This was done with a view to binding the entire volume, once completed, and to compiling a combined index to all its component parts. When binding the entire volume the pages marked with Roman numerals may be omitted.

PLAN OF FLORA OF SOUTHERN AFRICA

CRYPTOGAMS

Vol. 9 (Published 1978):   *Charophyta*

FLOWERING PLANTS

INTRODUCTORY VOLUMES

The genera of Southern African flowering plants:
    Vol. 1 (Published 1975):   *Dicotyledons*
    Vol. 2 (Published 1976):   *Monocotyledons*

OTHER VOLUMES

Vol. 1 (Published 1966):   *Stangeriaceae, Zamiaceae, Podocarpaceae, Pinaceae*, Cupressaceae, Welwitschiaceae, Typhaceae, Zosteraceae, Potamogetonaceae, Ruppiaceae, Zannichelliaceae, Najadaceae, Aponogetonaceae, Juncaginaceae, Alismataceae, Hydrocharitaceae*

Vol. 2: Poaceae

Vol. 3: Cyperaceae, Arecaceae, Araceae, Lemnaceae, Flagellariaceae

Vol. 4: Restionaceae, Mayacaceae, Xyridaceae, Eriocaulaceae, Commelinaceae, Pontederiaceae, Juncaceae

Vol. 5: Liliaceae, Agavaceae

Vol. 6: Haemodoraceae, Amaryllidaceae, Hypoxidaceae, Tecophilaeaceae, Velloziaceae, Dioscoreaceae

Vol. 7: Iridaceae

Vol. 8: Musaceae, Strelitziaceae, Zingiberaceae, Cannaceae*, Burmanniaceae, Orchidaceae

Vol. 9: Casuarinaceae*, Piperaceae, Salicaceae, Myricaceae, Fagaceae*, Ulmaceae, Moraceae, Cannabaceae*, Urticaceae, Proteaceae

Vol. 10:

  Part 1 (Published 1979): *Loranthaceae, Viscaceae*

      Santalaceae, Grubbiaceae, Opiliaceae, Olacaceae, Balanophoraceae, Aristolochiaceae, Rafflesiaceae, Hydnoraceae, Polygonaceae, Chenopodiaceae, Amaranthaceae, Nyctaginaceae

Vol. 11: Phytolaccaceae, Aizoaceae, Mesembryanthemaceae

Vol. 12: Portulacaceae, Basellaceae, Caryophyllaceae, Illecebraceae, Cabombaceae, Nymphaeaceae, Ceratophyllaceae, Ranunculaceae, Menispermaceae, Annonaceae, Trimeniaceae, Lauraceae, Hernandiaceae, Papaveraceae, Fumariaceae

Vol. 13: (Published 1970): *Brassicaceae, Capparaceae, Resedaceae, Moringaceae, Droseraceae, Roridulaceae, Podostemaceae, Hydrostachyaceae*

Vol. 14: Crassulaceae

---

* Alien families are marked with an asterisk

vi

LORANTHACEAE

by D. Wiens* and H. R. Tölken**

Shrubby, brittle, perennial aerial hemiparasites of other dicotyledons, glabrous or variously pubescent, often with swollen nodes. *Leaves* opposite, semi-opposite, alternate or whorled, simple, entire, often coriaceous, predominantly evergreen (rarely deciduous), exstipulate. *Flowers* bisexual, dichlamydeous, usually large and brightly coloured with yellows and reds, pollinated mostly by birds. *Calyx* often reduced to a low rim (calyculus), persistent on fruit. *Corolla* choripetalous or gamopetalous, actinomorphic, or zygomorphic by occurrence of a unilateral split. *Stamens* epipetalous, mostly basi-fixed, as many as the petals. *Anthers* mostly 2-loculate or sometimes multiloculate by presence of numerous transverse septa. *Pollen* mostly trilobiate. *Ovary* inferior, sometimes with disc, uni- or plurilocular, typically without distinct placenta. *Style* simple; stigma mostly capitate. *Ovules* absent, embryo sacs formed at base of ovary (mamelon). *Fruit* a berry or drupe, baccate (rarely dry), with a viscous layer developed outside vascular bundles. *Seeds* without testa; embryo mostly cylindrical with 2–6 cotyledons. *Basic chromosome number* x=12.

Characters not applicable in Southern Africa: Rarely root parasitic terrestrial shrubs or small trees; rarely dioecious.

A family of approximately 65 genera and about 900 species; widely distributed through the tropical and south temperate regions of the world.

Dyer (Gen., 1975) still included Viscaceae under Loranthaceae. He also referred all species here placed in the family Loranthaceae to the genus *Loranthus*. In the light of recent research, especially the pollination mechanism of the flowers, subdivision of the genus *Loranthus* sens. lat. into smaller genera along the lines suggested by earlier taxonomists, appears justified. *Loranthus* sens. strict. comprises only the single species *L. europaeus* Jacq. (For further discussion and synonymy see Barlow in Proc. Linn. Soc. N.S.W. 89: 268–272; 1969 and Kruijt in Brittonia 20: 136–147; 1968).

The flowers of many Southern African Loranthaceae open explosively. Mature buds are usually triggered when sunbirds probe through small apertures which develop between the corolla lobes when the buds mature; however, other trigger systems exist. When the flower opens, the pollen is released in a single explosive cloud in the direction of the probe. This action dusts the forehead of the sunbird with pollen (for further details see M. S. Evans in Nature 51: 235; 1895). The pollen is dispersed by an instantaneous coiling or inflexing of the filaments. Explosive anthesis is thus directly correlated with the occurrence of coiled or inflexed filaments. Furthermore, in genera with explosive anthesis and zygomorphic corollas (i.e., those with a "V"-shaped unilateral split), the split in the corolla also develops at the point of the original probe. When plants with mature floral buds are dried, turgor is lost and the buds may open gradually upon desiccation. Anthesis is thus not explosive and the flowers open only partially. In such cases the unilateral split may not develop normally and the flowers will appear to be regular. Special care should thus be exercised to use flowers which have opened spontaneously.

* Department of Biology, University of Utah, Salt Lake City, Utah 84112, U.S.A.
** Formerly Botanical Research Institute, Pretoria. Present address: State Herbarium, North Terrace, Adelaide, South Australia 5000, Australia.

Published: 1979.

1 Flowers pentamerous:

  2 Corolla sympetalous, forming at least a short, but more often a conspicious basal tube, essentially straight above the often swollen base:

    3 Flowers bilaterally symmetrical following anthesis (bearing a single "V"-shaped unilateral split which is sometimes obscured in pressed specimens):

      4 Filaments following anthesis coiled or inflexed 90 degrees or more, 5 mm or more long; styles essentially glabrous:

        5 Style gradually thickened from above the middle and abruptly constricted into a conspicuous neck below the enlarged and rounded stigma (skittle-shaped) . . . . .1. **Tapinanthus**

        5 Style with essentially the same diameter throughout its length (filiform):

          6 Corolla lobes distal from point of filament attachment, usually much shorter or occasionally as long as the rest of the corolla; corolla tube below split (i.e. closed portion) c. 10 mm or longer:

            7 Corolla tube without conspicuous basal swelling, if slightly expanded basally, then never abruptly constricted, enlarging laterally to at least 3 mm or more in diam. . . . . .1. **Tapinanthus**

            7 Corolla tube with conspicuous basal swelling, then abruptly and very narrowly constricted (c. 0,5 mm) just above swelling, thereafter enlarging laterally to c. 1,5 mm . . . . .2. **Tieghemia**

          6 Corolla lobes distal from point of filament attachment, at least twice the length of rest of corolla; corolla tube below split (i.e. closed portion) c. 7 mm or less long:

            8 Corolla reddish orange (mature buds black-tipped), lobes reflexed into coils; anthers typically breaking from filaments at anthesis; inflorescence primarily umbellate, but occasionally an apically crowded raceme . . . . .3.   **Moquinella**

            8 Corolla greenish yellow, lobes erect to slightly recurved, never coiled; anthers mostly persistent following anthesis; inflorescence umbellate . . . . .4. **Odontella**

      4 Filaments following anthesis essentially erect but often curving slightly inward, c. 3 mm long; style conspicuously tomentose in lower half . . . . . . .5. **Vanwykia**

    3 Flower radially symmetrical following anthesis (not producing a "V"-shaped split):

      9 Corolla bearing conspicuously long (3–5 mm), white, silky, ascending hairs; young leaves generally light to densely stellate-pubescent; inflorescence a head; anthers typically breaking from filaments at anthesis . . . .6. **Erianthemum**

      9 Corolla, leaves, and inflorescence essentially glabrous; inflorescence umbellate, anthers mostly persistent following anthesis:

        10 Filaments curving gently outward to 90 degrees following anthesis, essentially isodiametric; style broadly recurved c. 180 degrees near apex; corolla tube expanding slightly upward, 35–40 mm long, 3–5 mm wide, greenish yellow . . . . .7. **Pedistylis**

        10 Filaments tightly coiled following anthesis, coiled portion distinctly thicker and wider than rest of filament; corolla tube 15–20 mm long, 1–2 mm wide, crimson and yellow . . . . .8. **Actinanthella**

2 Corolla choripetalous (but the petals connivent and superficially sympetalous); buds and open flowers strongly curved or bow-shaped....................9. **Plicosepalus**

1 Flower tetramerous (rarely pentamerous):

11 Inflorescence a raceme; corolla choripetalous, (but petals basally connivent and superficially sympetalous), glabrous, radially symmetrical ....10. **Helixanthera**

11 Inflorescence a single to few-flowered, sessile umbel; corolla canescent or tomentose, sympetalous, bilaterally symmetrical (bearing a unilateral "V"-shaped split) ....11. **Septulina**

2074a   1. TAPINANTHUS*

Tapinanthus (*Blume*) *Reichb.*, Repert. Herb. 73 (1841); v. Tieghem in Bull. Soc. bot. Fr. 42: 267 (1895); Danser in Verh. K. Akad. Wet., sect. 2, 29, 6: 107 (1933); Balle in F.S.W.A. 22: 10 (1968); Stearn in Taxon 17: 157 (1968). Type species: *T. sessilifolius* (Beauv.) v. Tieghem.

Loranthus sect. *Tapinanthus* Blume, Fl. Jav., Loranth. 15 (1830); Engl. in Pflanzenfam. ed. 1, 3, 1: 187 (1894); in Bot. Jb. 20: 107 (1894); Benth. & Hook.f., Gen. Pl. 3, 1: 210 (1880), pro parte; Sprague in Kew Bull. 1914: 367 (1914); 1915: 70 (1915). *L.* subgen. *Tapinanthus* (Blume) Engl. in Pflanzenfam. ed. 1, Nachtr. 1: 132 (1897).

Lichtensteinia Wendl., Coll. Pl. 2: 5 (1808) (nom. rejic.), non Cham. & Schlechtd. (1826). *Loranthus* sect. *Lichtensteinia* (Wendl.) Blume, Fl. Jav., Loranth. 14 (1830).

Acranthemum v. Tieghem in Bull. Soc. bot. Fr. 42: 255 (1895). *L.* sect. *Dendrophthoe* (Mart.) Engl. in Pflanzenfam. ed. 1, 3, 1: 186 (1894), pro parte; Engl. & Krause in Pflanzenfam. ed. 2, 16b: 152 (1935), pro parte quoad 'Gruppe' *Acranthemum*. *L.* subgen. *Dendrophthoe* (Mart.) Engl. in Pflanzenfam. ed. 1, Nachtr. 1: 131 (1897), pro parte quoad 'Gruppe' *Acranthemum*.

Agelanthus v. Tieghem in Bull. Soc. bot. Fr. 42: 246 (1895); Balle in Webbia 11: 583 (1955), emend.; in Mitt. bot. StSamml., Münch. 7: 154 (1968); in F.S.W.A. 22: 3 (1968). *Loranthus* subgen. *Agelanthus* (v. Tieghem) Balle in F.C.B. 1: 309 (1948).

Phragmanthera v. Tieghem in Bull. Soc. bot. Fr. 42: 261 (1895); Balle in Webbia 11: 583 (1955); in Mitt. bot. StSamml., Münch. 7: 164 (1968); in F.S.W.A. 22: 5 (1968).

L. sect. *Constrictiflori* Sprague in F.T.A. 6, 1: 268 (1910), pro parte; Engl. & Krause in Pflanzenfam. ed. 2, 16b: 166 (1935).

Shrubs of varying sizes from approximately 0,5–2 m or higher, glabrous to densely and variously pubescent, stems usually glabrate with age, often with swollen, floriferous nodes. *Haustorium* with a single primary penetrating organ. *Inflorescence* usually an axillary umbel or sometimes a head, often fascicled, occasionally flowers solitary through reduction of the peduncles. *Flowers* 5-merous, gamopetalous, bilaterally symmetrical by the presence of a unilateral, "V"-shaped split of varying length. *Corolla* with a conspicuously swollen base, or tube cylindrical, lobes erect or reflexed, glabrous or variously pubescent, mostly yellow, red or combinations thereof. *Filaments* coiled or involutely curved at anthesis as the result of explosive opening of flower, anthers with or without basal tooth or ledge. *Style* filiform or upper half thickened gradually from middle, then abruptly constricted into a neck below stigma.

The largest and central genus of African Loranthaceae with perhaps 200 species widely distributed throughout the continent except in the central and western Sahara and most of North Africa.

1 Corolla tube conspicuously swollen at base, then abruptly and narrowly constricted, followed by a gradual expansion upward; style gradually thickened upward from near the middle, but abruptly narrowed below stigma (skittle-shaped); filaments bearing a small (c. 1,0 mm or shorter) tooth or ledge below anther:

2 Corolla lobes reflexed:

3 Corolla glabrous:

4 Leaves with a minute, usually dark reddish margin, especially when young, chartaceous to lightly coriaceous with age; clavate portion of bud apex at maturity reddish, with dark purplish ribs along sutures of corolla lobes; umbel mostly 3–4-flowered, usually densely fascicled at nodes of older, thick (c. 10 mm) leafless stems ....1. *T. rubromarginatus*

---

* *Tapinanthus* (Blume) Reichb. (1841) proposed for conservation against *Tapeinanthus* Herbert (1837).

4 Leaves with a minute, usually dull, whitish margin, moderately to highly coriaceous with age; clavate portion of bud apex at maturity pale green, with dull, whitish ribs along sutures of corolla lobes; umbel mostly 2-flowered, more often associated with younger, leafy stems, if on older branches, then usually not in dense fascicles ....2. *T. forbesii*

3 Corolla lightly puberulent (lens may be necessary) to densely pubescent:

5 Leaves at maturity shortly petiolate (2–3 mm), never cordate, glabrous:

6 Corolla predominantly purple, villous-pilose pubescent ....3. *T. leendertziae*
6 Corolla predominantly pale pink, usually variegated with whitish spots, puberulent ....4. *T. oleifolius*
5 Leaves at maturity sessile, cordate-amplexicaul, velutinous ....5. *T. mollissimus*

2 Corolla lobes erect:

7 Corolla glabrous or variously whitish pubescent, never conspicuously brownish tomentose over entire surface:

8 Petioles and young stems of flowering shoots short, rusty-pubescent:

9 Leaves orbicular, oblong, ovoid (never more than 1,5 times longer than wide), usually 30–40 mm long; filaments red ....6. *T. ceciliae*
9 Leaves lanceolate, sometimes becoming falcate with age, c. 70–100 mm long, 10–25 mm wide; filaments same colour as inside of corolla tube.............................7. *T. carsonii*

8 Petioles and young flowering shoots essentially glabrous (rarely-white puberulent):

10 Leaves without obvious lateral veins, linear-lanceolate, usually less than 10 mm wide ....8. *T. sambesiacus*
10 Leaves with conspicuous lateral veins, lanceolate to ovate-orbicular, usually more than 15 mm wide:

11 Flowers crowded on short, stout lateral shoots 20–60 × 3–4 mm, these with internodes c. 10 mm or less long; corolla tube with split less than half as long as closed basal portion; leaves markedly thick and succulent ....9. *T. crassifolius*
11 Flowers on rather elongated, thin principal branches 1–2 mm wide with internodes usually over 15 mm long; corolla tube with split about as long as closed portion (±2–3 mm); leaves chartaceous-coriaceous:

12 Corolla glabrous, basal swelling mostly rounded, but truncated (or at least terminating abruptly) below constriction of corolla tube; ovaries usually with slight constriction below calyx rim ....10. *T. kraussianus*
12 Corolla lightly puberulous, basal swelling mostly oblong and tapering gradually into constriction above swelling; ovary without slight constriction below calyx rim ....11. *T. prunifolius*
7 Corolla conspicuously brownish tomentose over entire surface ....12. *T. terminaliae*

1 Corolla tube not conspicuously swollen at base, if sometimes slightly expanded basally, then never abruptly constricted above base; style filiform (±skittle-shaped in *T. natalitius* subsp. *zeyheri*); filaments without a tooth or ledge below anther:

13 Flowers terminating short, leafy, spur-branches 10–40 mm long at anthesis; bud apex at maturity beaked to uncinate:

14 Spur-branches, petioles, and pedicels pubescent to puberulent (rarely glabrous); style contracted for 6–8 mm below stigma (± skittle-shaped); corolla base 3–5 mm wide; berry c. 10 mm wide or more ....13. *T. natalitius*
14 Plant glabrous; style filiform; corolla base 1–2 mm wide; berry c. 5 mm wide ....14. *T. gracilis*
13 Flowers either axillary on younger shoots, borne at nodes of often leafless olders tems, or associated with sessile, leafy fascicles, never terminal on spur-branches; bud apex at maturity rounded to acute:

15 Corolla (and plants generally) essentially glabrous; calyx tubular, c. 2–3 mm long, minutely toothed; anthers not transversely septate:

16 Lobes of corolla erect, 16–18 mm long, about as long as tube; leaves at maturity 10–20 × 5–10 mm ....15. *T. discolor*
16 Lobes of corolla slightly recurved at point of filament attachment, 10–12 mm long, about half as long as tube; leaves generally 25–40 × 15–20 mm ....16. *T. lugardii*
15 Corolla (and plants generally) variously but conspicuously pubescent; calyx essentially reduced to a rim; anthers transversely septate:

17 Flowers and leaves uniformly grey-tomentose; corolla tube in bud essentially cylindrical or only slightly broadened distally, apex not winged; lobes of corolla linear or only slightly broadened at apex; leaves and flowers in dense fascicles, often on older stems; blades elongate, 10–15 mm long, about twice as long as broad; petioles less than 1 mm wide, approximately one third to one half as long as blade ....17. *T. guerichii*

17 Flowers and young leaves lanate or canescent, older leaves becoming less pubescent; corolla tube in bud apically elliptic to clavate and minutely but distinctly winged at the sutures (wings sometimes obscured by pubescence); lobes of corolla spathulate; leaves and flowers often crowded at shortened internodes of new shoots; blades broadened, not twice as long as broad, 20–40 mm long; petioles over 1 mm wide, about one fifth or less the length of the blade:

18 Flowers canescent, often more densely so at base; bracts inconspicuous, 1–2 mm long, not exceeding ovary; anthers c. 5 mm long ....18. *T. cinereus*

18 Flowers with long, spreading hairs; bracts becoming large and foliaceous, 10–15 mm long; anthers 1–2 mm long ....19. *T. glaucocarpus*

1. Tapinanthus rubromarginatus (*Engl.*)

Danser in Verh. K. Akad. Wet., sect. 2, 29, 6: 119 (1933). Type: Transvaal, Magaliesberg near Buffelspoort, *Engler* 2837a (K!).

Loranthus rubromarginatus Engl. in Bot. Jb. 40: 535 (1908); Sprague in F.C. 5, 2: 116 (1915); Burtt Davy, Fl. Transv. 465 (1932); Letty in Wild Flow. Transv. pl. 61, 3 (1962); Ross, Fl. Natal 152 (1972).

L. glabriflorus Conrath in Kew Bull. 1908: 226 (1908). Type: Transvaal, near Witpoortje, *Conrath* 331 (K, holo.!).

Relatively large shrubs up to 1 m or more high, glabrous, densely lenticelled. *Stems* thick and stout, up to 10 mm thick. *Leaves* in crowded fascicles, often on swollen nodes of older branches, deciduous in winter, elliptic to elliptic-oblong (20–) 30–40 (–50) × 10–15 mm, chartaceous to lightly coriaceous with age, penninerved; petioles 3–5 mm long. *Inflorescence:* umbels mostly densely clustered on swollen nodes of older stems, mostly 2–4-flowered, subsessile to shortly (1–2 mm long) pedunculate. *Corolla* with conspicuous, often oblong-rounded, basal swelling, 40–50 mm long, tube split 14–16 mm below lobes, dark red to purplish at base and apex, often variegated with whitish spots; in mature buds the clavate apex with small but conspicuous purplish wings at petal sutures; lobes reflexed near middle. *Filaments* with a tooth below anther. *Style* constricted below stigma. *Berries* rounded, c. 10 mm long, reddish. *Flowering* September through November; $n=9$. Fig. 1.

Parasitic on species of *Acacia, Chrysophyllum, Dichrostachys, Dombeya, Faurea, Populus, Protea, Prunus,* occurring throughout the Transvaal and in Swaziland and northern Natal (Map 1).

Vouchers: *Acocks* 23347; *Galpin* 10845; 11564; *Stauffer & Scheepers* 5245.

2. Tapinanthus forbesii (*Sprague*) *Wiens*

in Bothalia 12: 423 (1978). Type: Mozambique, Delagoa Bay, *Forbes* s.n. (K, lecto.!).

Loranthus oleifolius (Wendl.) Cham. & Schlechtd. var. *forbesii* Sprague in F.C. 5, 2: 118 (1915).

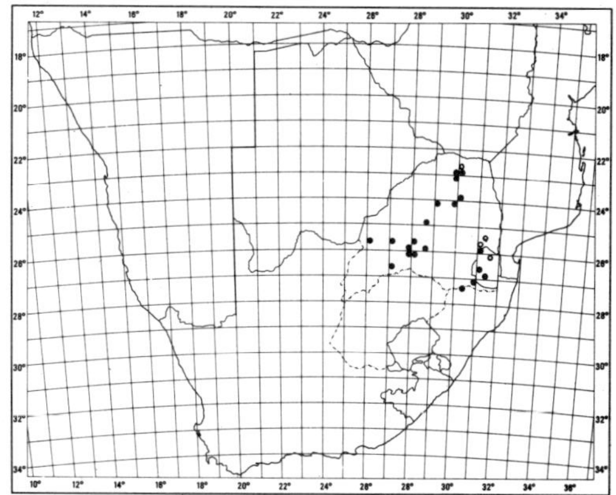

MAP 1.—●**Tapinanthus rubromarginatus**
○**T. forbesii**

Closely related to *T. oleifolius* and *T. rubromarginatus*. Distinct from *T. oleifolius* by a glabrous corolla and slightly winged sutures of clavate apical portion of mature buds, and differentiated from *T. rubromarginatus* by the dull pink corolla with greenish apex, and the oblong ovary (in *T. rubromarginatus* the ovary is about as long as wide). Fig. 1.

Parasitic on species of *Acacia* and *Sterculia*. Occurs in the northern and eastern Transvaal and in Mozambique (Map 1).

Vouchers: *Marais 273; Van der Schijff 1124; Wiens 5322* (a).

Although Sprague considered this species to be an element of *T. oleifolius*, which it does resemble, *T. forbesii* is morphologically isolated from that species. It is also ecologically differentiated by its occurrence in the lowveld of the northern and eastern Transvaal, north-eastern Swaziland, adjoining Mozambique, and probably Rhodesia.

3. Tapinanthus leendertziae (*Sprague*)

Wiens in Bothalia 12: 423 (1978). Type: Transvaal, near Potgietersrus, *Leendertz* 1142 (K, holo.!; PRE!).

Loranthus oleifolius (Wendl.) Cham. & Schlechtd. var. *leendertziae* Sprague in F.C. 5, 2: 118 (1915); Burtt Davy, Fl. Transv. 466 (1932).

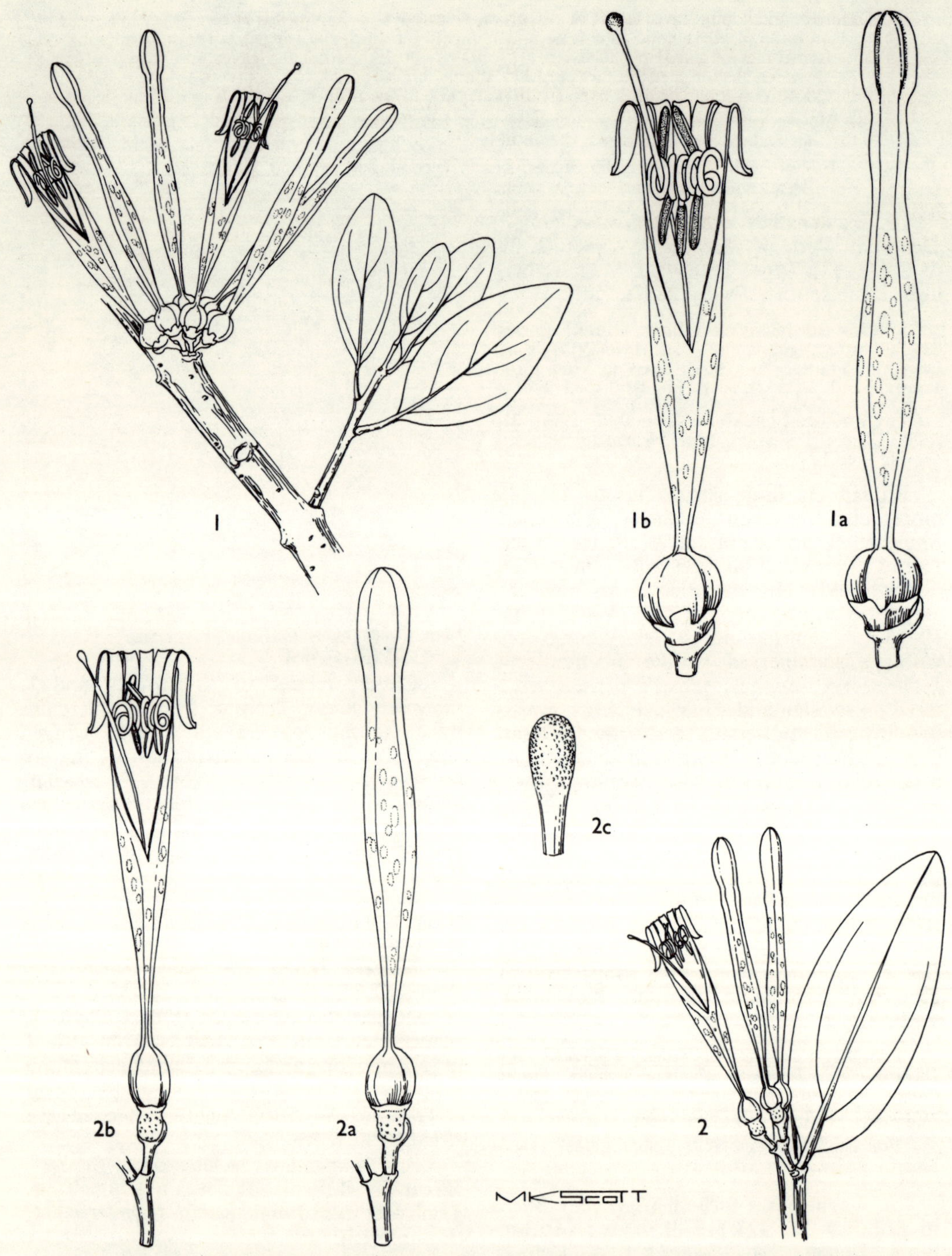

FIG. 1.—1, **Tapinanthus rubromarginatus,** flowering twig, ×1; 1a, mature bud, ×2; 1b, flower, ×2 (*Sidey* 2377). 2, **T. forbesii,** flowering twig, ×1; 2a, mature bud, ×2; 2b, flower, ×2; 2c, stigma, ×6 (*Wiens* 5321).

Relatively large shrubs over 1 m high. *Stems* rather thin, puberulent when young, glabrate with age. *Leaves* opposite-subopposite, occasionally fascicled on older stems, glabrous, mostly lanceolate-oblong, 30–40 × 10–20 mm, somewhat chartaceous, conspicuously penninerved; petioles 2–4 mm long, usually puberulent. *Inflorescence:* umbels sessile-subsessile, 4–6-flowered, mostly solitary in axils; pedicels and bracts pubescent. *Corolla* with conspicuous, rounded-oblong basal swelling, pilose, base green, tube reddish-purple, apex green, 35–40 mm long, tube split 10–12 mm below lobes; lobes reflexed. *Filaments* with a small tooth below anthers. *Style* constricted below stigma. *Berries* ellipsoid, 10–12 mm long, red, slightly warty. *Flowering* from April through September and probably also in other months; $n=9$. Fig. 2.

Parasitic primarily on species of *Acacia*, but also *Euclea*, *Ochna*, *Peltophorum* and *Rhus*. Occurs throughout the Transvaal and the northern Orange Free State, with an apparently disjunct population in northern Natal (Map 2).

Vouchers: *Codd* 2221; *Schlechter* 4291; *Werdermann & Oberdieck* 1633.

Although considered by Sprague to be a variety of *T. oleifolius*, *T. leendertziae* is quite distinct and appears to have little in common with *T. oleifolius*. Sprague mentioned the occurrence of intermediate collections but field studies in the northern Transvaal did not reveal their presence.

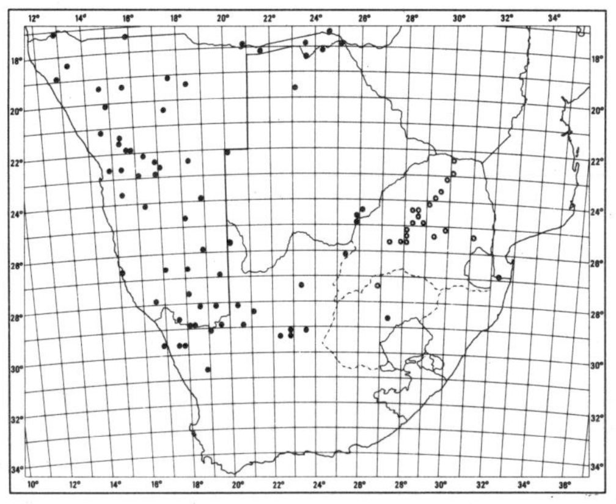

MAP 2.—○ **Tapinanthus leendertziae**
● **T. oleifolius**

4. **Tapinanthus oleifolius** (*Wendl.*) *Danser* in Verh. K. Akad. Wet., sect. 2, 29, 6: 117 (1933); Balle in Mitt. bot. StSamml., Münch. 7: 182 (1968); in F.S.W.A. 22: 11 (1969). Type: Cape, near Prieska, *Lichtenstein* s.n. in Herb. Wendland (GOET, holo.; PRE, photo.!).

Lichtensteinia oleifolia Wendl., Coll. Pl. 2: 5, t.39 (1808), as *oleaefolia*. *Loranthus oleifolius* (Wendl.) Cham. & Schlechtd. in Linnaea 3: 209 (1828), as *oleaefolius*; DC., Prodr. 4: 304 (1830), do; Harv. in F.C. 2: 576 (1862), do; Sprague in Kew Bull. 1914: 359 (1914), do; in F.C. 5,2: 117 (1915), do. *Scurrula oleifolia* (Wendl.) G. Don, Gen. Hist. 3: 423 (1834), as *oleaefolius*.

L. speciosus F.G. Dietr., Lexic. Gaertn., Nachtr. 4: 473 (1818), nom. superfl. *Lichtensteinia speciosa* (F.G. Dietr.) v. Tieghem in Bull. Soc. bot. Fr. 42: 254 (1895). Type: same as for *Lichtensteinia oleifolia*.

L. lichtensteinii Willd. ex Cham. & Schlechtd. in Linnaea 3: 209 (1828), in syn.

L. meyeri Presl, Bot. Bemerk. 76 (1844), nom. nud.

L. namaquensis Harv. in F.C. 2: 577 (1862); Sprague in F.T.A. 6, 1: 361 (1910). *Tapinanthus namaquensis* (Harv.) v. Tieghem in Bull. Soc. bot. Fr. 42: 267 (1895). Syntypes: Cape, Groenrivier, *Drège* s.n. (K!); near Verleptpram, *Drège* s.n. (K!); on Gariep, *Drège* s.n.; Namaqualand, *Wyley* s.n.; Modderfontein, *Whitehead* s.n.

L. namaquensis Harv. var. *ligustrifolius* Engl. in Bot. Jb. 20: 120 (1894); Hiern, Cat. Afr. Pl. Welw. 1, 4: 932 (1900). Syntypes: Angola, Benguella, on Bero River, *Welwitsch* 4858 (B; PRE, photo.!); banks of Maiombo River, *Welwitsch* 4860.

L. meyeri Presl var. *inachabensis* Engl. in Bot. Jb. 40: 535 (1908), nom. invalid. Type: South West Africa, Inachab, *Dinter* 914 (B, holo.; PRE, photo.!).

L. oleifolius (Wendl.) Cham. & Schlechtd. var. *luteus* Neusser in Mitt. bot. StSamml., Münch. 1: 339 (1953). Syntypes: South West Africa, Swakoptal, *Walter* 1288 (M!); Swakop to Palmenhorst, *Schönfelder* s.n.

Moderately large shrubs up to 1 m or more high. *Stems* puberulent when young, glabrate with age, often buff or brownish. *Leaves* subopposite to alternate (scattered), mostly ovate-elliptic but highly variable, 30–45 × 10–20 mm, coriaceous; petioles c. 3 mm long to subsessile. *Inflorescence:* umbels mostly axillary, solitary, 3–4-flowered, minutely puberulent; peduncles from absent up to 2 mm long and approximately equalling the pedicels. *Corolla* with conspicuous basal swelling, 35–40 mm long, tube split 10–12 mm below lobes, dull pink with whitish, irregular variegations, lightly puberulent, light green at apex and base; lobes reflexed. *Filaments* with small tooth below anther. *Style* constricted below stigma. *Berries* ellipsoid, 10–12 mm long, reddish orange. *Flowering* probably throughout the year; $n=9$. Fig. 2.

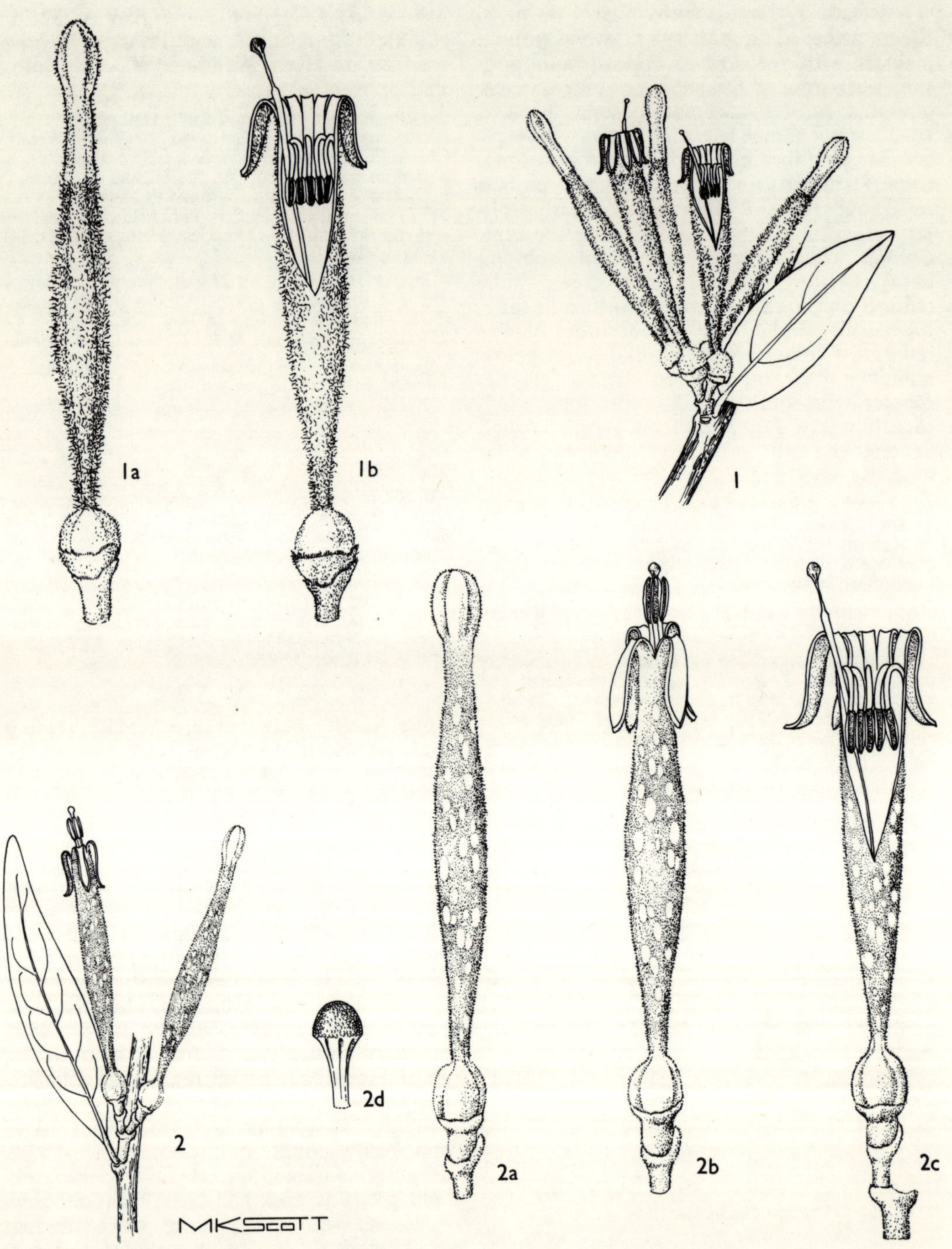

FIG. 2.—1, **Tapinanthus leendertziae**, flowering twig, ×1; 1a, mature bud, ×2; 1b, flower, ×2 (*Breyer* in TRV 13855). 2, **T. oleifolius**, flowering twig, ×1; 2a, mature bud, ×2; 2b, flower in first stage, ×2; 2c, opened flower, ×2; 2d, stigma, ×6 (*Taylor* 8438).

Parasitic on numerous and diverse hosts, primarily *Acacia*, but also on species of *Aloe, Citrus, Combretum, Cotyledon, Diospyros, Maytenus, Melianthus, Parkinsonia, Rhus, Terminalia* and *Ziziphus*. Widespread throughout South West Africa, much of Botswana, the western Transvaal and the central Orange Free State (Map 2).

Vouchers: *Galpin* 9450; *Marloth* 12385; *Smith* 2378.

5. Tapinanthus mollissimus (*Engl.*) *Danser*

in Verh. K. Akad. Wet., sect. 2, 29, 6: 116 (1933). Syntypes: Angola, between Bruco & Chao da Xella, *Welwitsch* 4877 (K!); near Monino, *Welwitsch* 4888 (K!).

Loranthus mollissimus Engl. in Bot. Jb. 20: 119 (1894); Hiern in Cat. Afr. Pl. Welw. 1, 4: 934 (1900); Sprague in F.T.A. 6, 1: 358 (1910).

Shrubs probably reaching 1 m or more high, mostly villous-pubescent, velutinous, older stems glabrate. *Leaves* opposite, sessile, ovate to occasionally rounded, 35–50 (–90) × 25–35 (–70) mm, conspicuously cordate-amplexicaul with age, venation conspicuous but lateral veins branching somewhat irregularly, heavily coriaceous. *Inflorescence*: heads in axillary fascicles, mostly 4-flowered; peduncles 1–2 mm long. *Corolla* with conspicuous basal swelling, velutinous, yellowish (with scarlet?), 30 mm long, tube split c. 12 mm below reflexed lobes. *Filaments* with small tooth below anthers. *Style* constricted below stigma. *Berries* unknown. *Flowering* in May and no doubt also in other months. Fig. 5.

Parasitic on *Ficus* in Southern Africa where it is known only from Etosha Pan National Park in South West Africa; also in Angola.

Voucher: *Le Roux* 1136.

A highly distinct species in Southern Africa; known in our region only from a single collection from Etosha Pan. The species is closely related, and possibly conspecific with *T. mechowii* (Engl.) v. Tieghem, an Angolan species.

6. Tapinanthus ceciliae (*N.E.Br.*) *Danser*

in Verh. K. Akad. Wet., sect. 2, 29, 6: 110 (1933), as *cecilae*. Type: Rhodesia, Bulawayo, *Cecil* 96 (K, holo.!).

Loranthus ceciliae N.E.Br. in Kew Bull. 1906: 168 (1906), as *cecilae*; Sprague in F.T.A. 6, 1:378 (1910), as *cecilae*.

Shrubs 1 m or more long, often pendulous with age. *Stems* rather elongated; young internodes densely, short, brownish-pubescent, brownish glabrate with age, densely lenticellate. *Leaves* subopposite-alternate, broadly ovate to rounded, 30–50 ×

25–35 mm, glabrous, usually glaucous, penninerved, veins raised below, less obvious above; petioles 8–12 mm long, mostly densely and shortly brownish pubescent. *Inflorescence*: umbels solitary or several in axils, mostly 8–13-flowered, densely crowded; peduncles almost absent or up to 2 mm long, slightly shorter than pedicels, both minutely whitish pubescent. *Corolla* with conspicuous basal swelling and erect lobes, orange-yellow, greenish yellow basally, 35–40 mm long, tube split 10–12 mm below lobes, whitish pubescent (especially at base). *Filaments* with small tooth below anthers. *Style* constricted below stigma. *Berries* unknown. *Flowering* in April and May and probably other times also. Fig. 3.

Apparently parasitic only on other misletoes including species of *Plicosepalus, Tapinanthus* (?) and *Viscum*. Known only from the northern and north-eastern Transvaal; also in Rhodesia.

Vouchers: *Wiens* 5325; *Wiens & Van Wyk* 5330; 5340.

The Southern African populations of this mistletoe are placed in *T. ceciliae* with uncertainty. The entire complex to which *T. ceciliae* belongs [i.e. *T. dichrous* (Engl.) Danser, *T. schweinfurthii* (Engl.) Danser, *T. blantyreanus* (Engl.) Danser] needs further critical study before its elements can be determined with accuracy. This mistletoe is one of the few Loranthaceae which are apparently parasitic only on other Loranthaceae and *Viscum*. Epiparasitism of this nature is more common in viscaceous mistletoes, but also occurs in *T. gracilis* and *T. kraussianus*. The ecological aspects of these epiparasitic mistletoes are virtually unstudied, and further work on this fascinating phenomenon should be undertaken.

7. Tapinanthus carsonii (*Bak. & Sprague*)

Danser in Verh. K. Akad. Wet., sect. 2, 29, 6: 109 (1933). Type: Zambia, Fwambo near Abercorn, *Carson* s.n. (K, holo.!).

Loranthus carsonii Bak. & Sprague in F.T.A. 6, 1: 376 (1910).

Shrubs up to 2 m or perhaps higher, pendulous with age. *Stems* rather elongated; young internodes short, brownish pubescent, glabrate and brown with age. *Leaves* opposite-subopposite, mostly falcate-lanceolate, 80–100 × 20–30 mm, thickly coriaceous, penninerved; petioles 10–12 mm long, often shortly brownish pubescent. *Inflorescence*: umbels solitary, axillary, mostly 8–12-flowered, densely crowded and recurved on pendulous branches; peduncles and pedicels 2–4 mm long, minutely whitish puberulent. *Corolla* with conspicuous basal swelling, greenish yellow with apical orange band, 35–40 mm

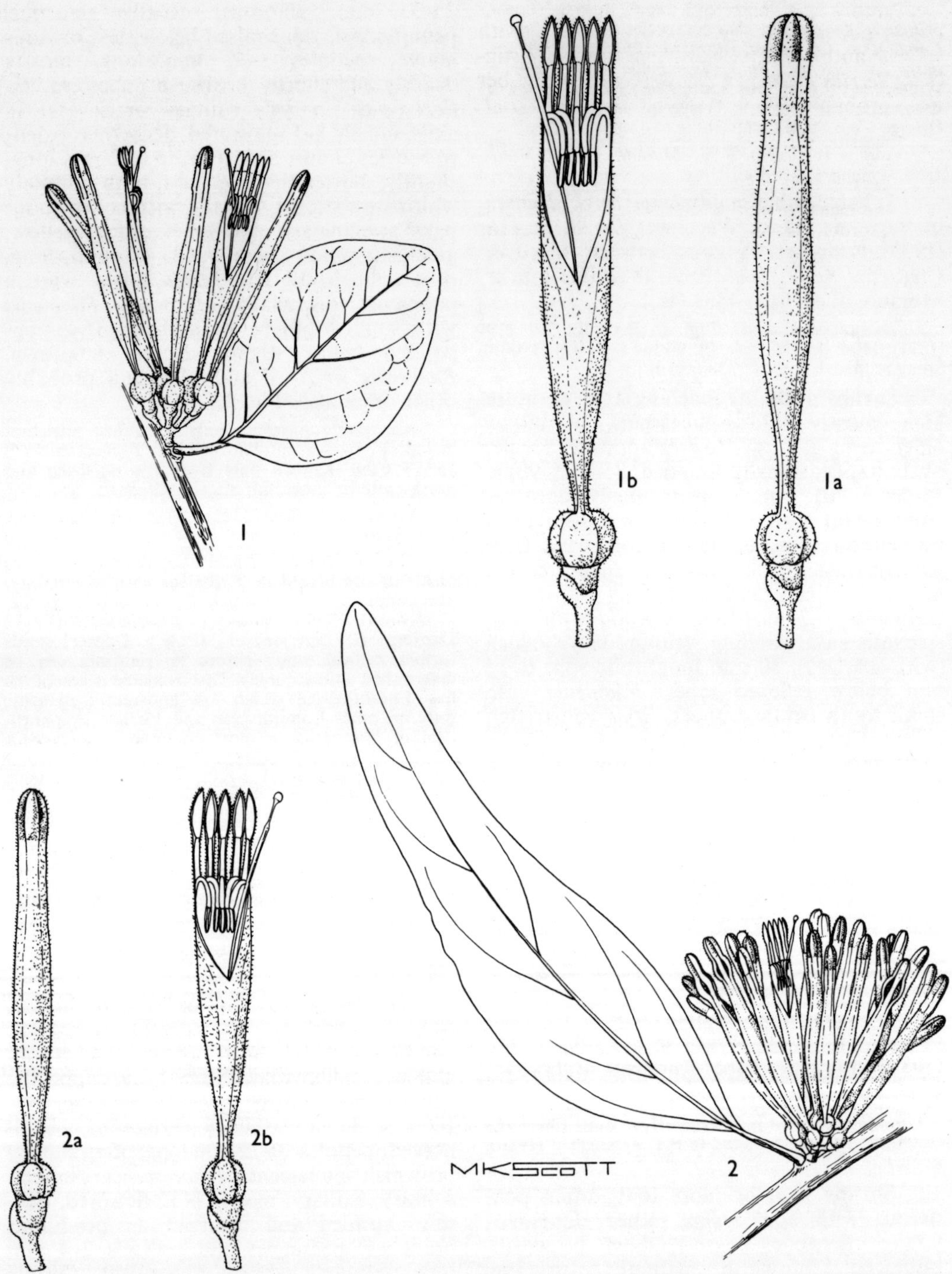

FIG. 3.—1, **Tapinanthus ceciliae**, flowering twig, ×1; 1a, mature bud, ×2; 1b, flower, ×2 (*Wiens* 5325). 2, **T. carsonii**, flowering twig, ×1; 2a, mature bud, ×2; 2b, flower, ×2 (*Edwards* 4310).

long, tube split 8–10 mm below lobes, shortly white-pubescent. *Filaments* with small tooth below anthers. *Style* constricted below stigma. *Berries* unknown. *Flowering* in November and probably also in other months. Fig. 3.

Parasitic on *Albizia* in the eastern Caprivi Strip.

Voucher: *Edwards* 4310.

The inclusion of this collection in *T. carsonii* is highly problematical. The type specimen appears similar to this collection, but no other material is apparently available and the geographical discontinuity between the type locality in Zambia and the Caprivi Strip is considerable. The patterns of variation and geographical distribution of *T. carsonii* must be understood before this population can be placed in the species with confidence.

The name *T. carsonii* is used here, but as the tropical species become better known it may have to be included in the synonomy of *T. dichrous* (Engl.) Danser, as in Balle (F.C.B. 1: 339; 1948).

8. **Tapinanthus sambesiacus** (*Engl. & Schinz*) *Danser* in Verh. K. Akad. Wet., sect. 2, 29, 6: 119 (1933). Type: Mozambique, Boruma, *Menyharth* 934 (Z, holo.!; B!).

Loranthus sambesiacus Engl. & Schinz in Denkschr. Akad. Wiss. Wien, Math.-naturwiss. Kl. 78: 409 (1906); Sprague in F.T.A. 6, 1: 370 (1910).

Shrubs probably less than 0,5 m high, glabrous. *Stems* relatively thin, greyish, lenticellate. *Leaves* subopposite-alternate, lanceolate-linear, 40–60 × 5–10 mm, becoming highly coriaceous with age; petioles 4–5 mm long, flattened abaxially. *Inflorescence:* umbels solitary, axillary, mostly 3-flowered; peduncles c. 2 mm long, approximately equalling pedicels. *Corolla* with conspicuous basal swelling, yellow-orange, 30–35 mm long, tube split 7–9 mm below lobes; lobes erect. *Filaments* with tooth below anther. *Style* constricted below stigma. *Berries* obovoid, 10–12 mm long. *Flowering* from December through February. Fig. 4.

Known only as a parasite on *Commiphora* species in the Transvaal north of the Soutpansberg and probably in adjoining Rhodesia.

Vouchers: *Pienaar* 356; *Theron* 2864; *Strey* 3498; *Van der Schijff* 5213.

The isotype (B) is marked *Menyharth* 939 but this is probably an error.

9. **Tapinanthus crassifolius** *Wiens* in Bothalia 12: 422 (1978). Type: Transvaal, Kruger National Park, Pafuri area, *Codd & Dyer* 4637 (PRE, holo.!; K!).

Densely branched shrubs to 1 m or higher, essentially glabrous. *Stems* relatively thick, succulent and beige-coloured with dark brown lenticels when young, grey with age. *Leaves* opposite-subopposite to scattered, borne on relatively short (50–100 mm), succulent lateral branches; blades succulent, mostly ovate to broadly lanceolate, variable in size (40–) 60–80 (–110) × 30–50 (–110) mm, often slightly falcate, penninerved; petioles 15–20 mm long, flattened abaxially. *Inflorescence:* umbels densely crowded on short, lateral, leaf-bearing branches, mostly 4-flowered, axillary and extra-axillary; peduncles 2–4 mm long, approximately equalling pedicels. *Corolla* with conspicuous oblong basal swelling and erect lobes, orange-yellow with red bands near apex, 40–45 mm long, tube split 7–9 mm below lobes. *Filaments* with small tooth below anther. *Style* constricted below stigma. *Berries* obovoid, 10–12 mm long, whitish orange. *Flowering* in winter. Fig. 4.

A parasite on *Sclerocarya caffra* Sond. Apparently restricted to the northern and north-eastern Transvaal; to be expected in adjoining Rhodesia and Mozambique.

Vouchers: *Codd & Dyer* 4637; *Wiens & Van Wyk* 5332; 5335; *Van Wyk* 4721.

10. **Tapinanthus kraussianus** (*Meisn.*) *v. Tieghem* in Bull. Soc. bot. Fr. 42: 257 (1895). Type: Natal, near Durban, *Krauss* 125 (K).

Loranthus kraussianus Meisn. in Hooker, Lond. J. Bot. 2: 539 (1843); Harv. in F.C. 2: 577 (1862); Sprague in F.C. 5, 2: 118 (1915).

Shrubs up to 1 m high, glabrous. *Branches* relatively thin, usually less than 5 mm thick. *Leaves* subopposite-alternate, mostly lanceolate to ovate-rounded, (30–) 50–70 × 15–25 mm, penninerved; petioles 5–10 mm long. *Inflorescence:* umbels solitary, axillary on leafy shoots mostly (4–) 6–8-flowered; peduncles 3–5 mm long, approximately equalling pedicels. *Corolla* with conspicuous, rather rounded, somewhat lobed, basal swelling, predominantly orange-reddish with deep orange bands near apex, greenish orange basally, 30–45 mm long, tube split 10–15 mm below lobes; lobes erect. *Filaments* with small tooth below anther. *Style* constricted below stigma. *Berries* obovoid, 10–12 mm long, pink. *Flowering* mostly from December through March, but occasionally also in other months.

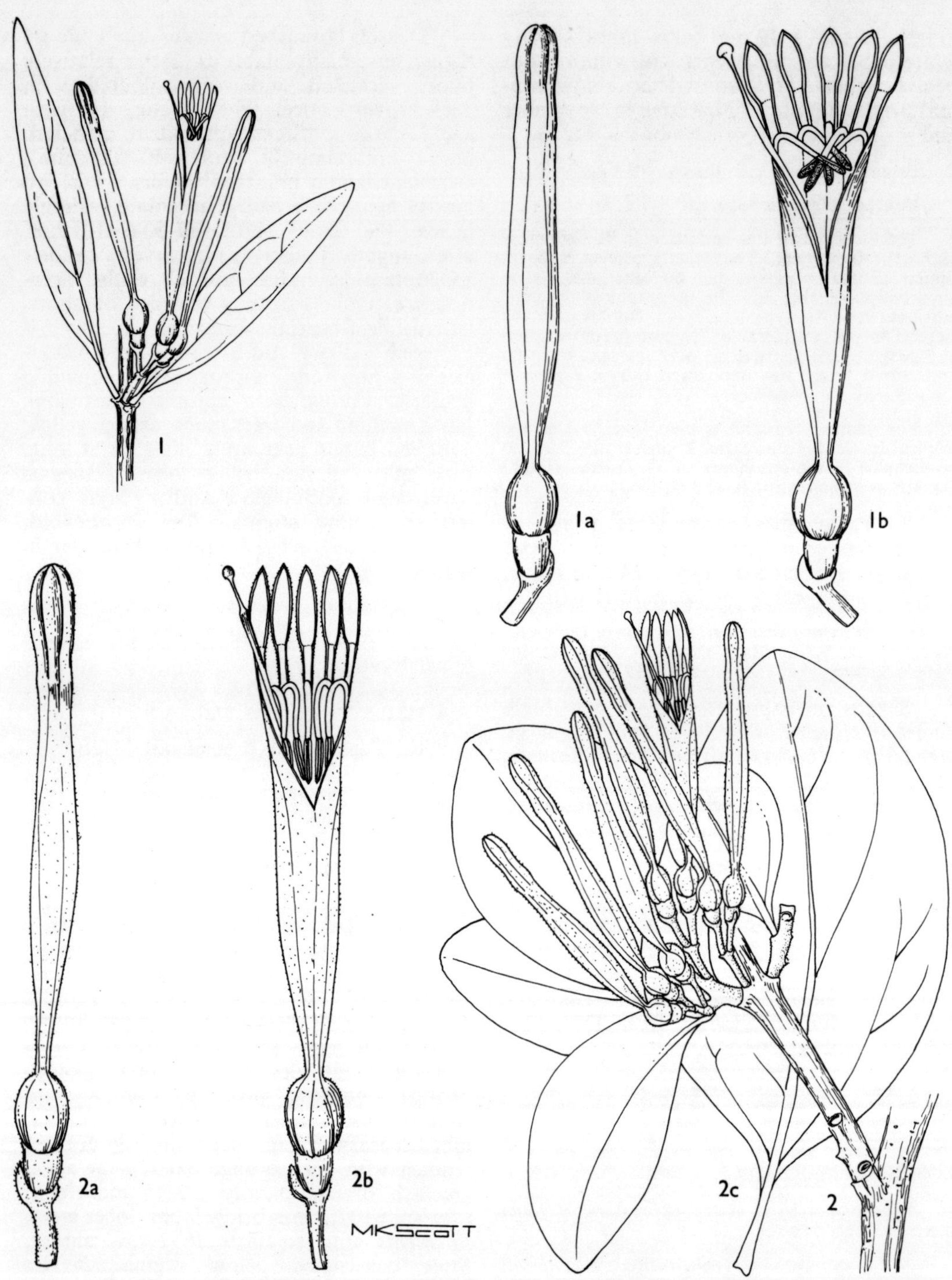

LORANTHACEAE

FIG. 4.—1, **Tapinanthus sambesiacus**, flowering twig, ×1; 1a, mature bud, ×2; 1b, flower, ×2 (*Pienaar* 356). 2, **T. crassifolius**, flowering twig, ×1; 2a, mature bud, ×2; 2b, flower, ×2 (*Codd & Dyer* 4637); 2c, typical leaf, ×1 (*Wiens* 5332).

Parasitic on diverse and numerous hosts including species of *Acacia, Bauhinia, Capparis, Celtis, Chaetachme, Combretum, Fluggia, Grewia, Prunus, Rhoicissus, Sapindus,* other *Tapinanthi, Turraea, Urera* and *Viscum.* Known from the south-eastern Transvaal through the Natal lowlands to Transkei, and probably the eastern Cape Province (Map 3).

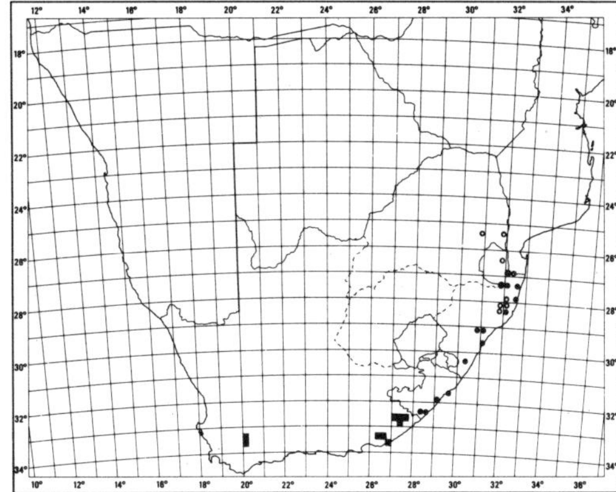

MAP 3.—●*Tapinanthus kraussianus* subsp. **kraussianus**
   ○T. **kraussianus** subsp. **transvaalensis**
   ■T. **prunifolius**

The species is highly variable and is divided into 2 subspecies originally defined by Sprague. The typical form occurs from central Natal southward, possibly to the eastern Cape Province. Subspecies *transvaalensis* occurs from the eastern Transvaal (lowveld) southward through the low elevations of Swaziland to central Natal. The region in central Natal, where the subspecies appear to intergrade, should be studied to define the distributional areas of these taxa and to determine if the recognition of 2 subspecies is the best taxonomic treatment of this highly variable species. (See also the discussion under *T. prunifolius*).

Branches greenish (darkened when dry); leaves broadened, mostly ovate-rounded, deep green; from approximately central Natal southward, possibly to the eastern Cape Province..............(a) subsp. *kraussianus*
Branches buff-brown; leaves elongate, mostly lanceolate, light grey-green, thickly coriaceous; from south-eastern Transvaal and Swaziland to central Natal ....
...................(b) subsp. *transvaalensis*

(a) subsp. **kraussianus.**

Loranthus kraussianus Meisn. in Hooker, Lond. J. Bot. 2: 539 (1843); Harv. in F.C. 2: 577 (1862); Wood & Evans, Natal Plants 1: 62, 1 pl. 76 (1899); Sprague in F.C. 5, 2: 118 (1915); Batten & Bokelmann, Wild Flow. E. Cape Prov. pl. 55, 1 (1966); Ross, Fl. Natal 152 (1972); Gibson, Wild Flow. Natal pl. 48, 4 (1975). *Tapinanthus kraussianus* (Meisn.) v. Tieghem in Bull. Soc. bot. Fr. 42: 257 (1895).

Distinguished from subspecies *transvaalensis* primarily by the characters mentioned in the key (Map 3).

Voucher: *Pegler* 292.

(b) subsp. **transvaalensis** (*Sprague*) *Wiens* in Bothalia 12: 423 (1978). Type: Transvaal, near Barberton, *Galpin* 879 (K, holo.!; PRE!).

Loranthus kraussianus Meisn. var. *transvaalensis* Sprague in F.C. 5, 2: 119 (1915); Burtt Davy, Fl. Transv. 466 (1932).

Distinguished from the typical subspecies primarily by the characters mentioned in the key and possibly also by having smaller flowers. Fig. 5; Map 3.

Vouchers: *Wiens* 5256; 5286.

11. **Tapinanthus prunifolius** (*E. Mey. ex Harv.*) *v. Tieghem* in Bull. Soc. bot. Fr. 42: 267 (1895); Danser in Verh. K. Akad. Wet., sect. 2, 29, 6: 118 (1933). Type: Cape, Glenfilling, *Drège* s.n. (K!).

Loranthus prunifolius E. Mey. ex Harv. in F.C. 2: 578 (1862); Sprague in F.C. 5, 2: 119 (1915).

L. kraussianus Meisn. var. *puberulus* Sprague in F.C. 5, 2: 119 (1915). Type: eastern Cape or western Transkei, *Flanagan* 25 (K, holo.!; PRE!).

Closely related to *T. kraussianus* from which it is most easily distinguished by the following characteristics: (1) the longer (4–8 mm), ellipsoid, basal swelling of the corolla which tapers gently into the constriction above the base; (2) the minutely puberulent flowers and inflorescences; and (3) the slight constriction of the ovary below the calyx rim. Fig. 5.

Parasitic on species of *Rhoicissus, Ficus,* and probably also other genera. Known only from the south-western and eastern Cape (Map 3).

Vouchers: *Acocks* 9368; 11016; *Galpin* 316; 2922; *Kirkman* 3084.

This species is possibly conspecific with *T. kraussianus.* Field studies on both taxa in southern Natal and the eastern Cape will be necessary to determine the morphological and geographical consistency of the characters. *Loranthus kraussianus* var. *puberulus* is tentatively placed here pending a more detailed field study of its relationships to both *T. kraussianus* and *T. prunifolius.* Very little material of either taxon is presently available for study.

12. **Tapinanthus terminaliae** (*Engl. & Gilg*) *Danser* in Verh. K. Akad. Wet., sect. 2, 29, 6: 120 (1933); Balle in Mitt. bot. StSamml., Münch. 7: 188 (1968); in F.S.W.A. 22: 11 (1969). Type: Angola, between Ungombekike and Kuito, *Baum* 519 (B, holo.!; PRE, photo.!).

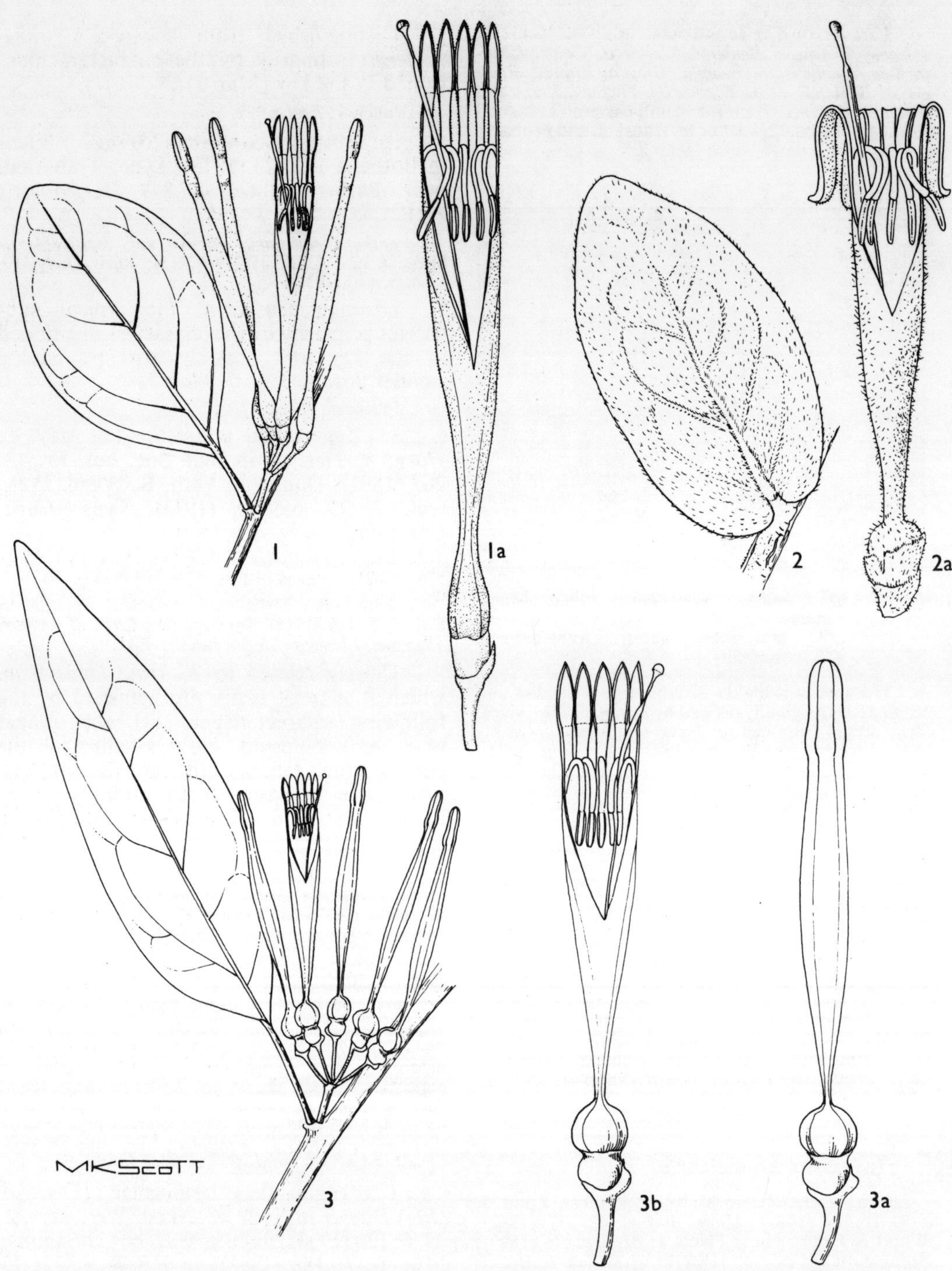

Fig. 5.—1, **Tapinanthus prunifolius**, flowering twig, ×1; 1a, flower, ×2 (*Acocks* 9368). 2, **T. mollissimus**, twig with leaf, ×1; 2a, flower, ×2 (*Le Roux* 1136). 3, **T. kraussianus** subsp. **transvaalensis**, flowering twig, ×1 (*Breyer* in TRV 17901); 3a, mature bud, ×2; 3b, flower, ×2 (*Wiens* 5256).

Loranthus terminaliae Engl. & Gilg in Warb. Kunene-Samb. Exped. 228 (1903); Sprague in F.T.A. 6, 1: 379 (1910).

L. villosiflorus Engl. in Bot. Jb. 20: 125 (1894), pro parte quoad *Welwitsch* 4890; Sprague in F.T.A. 6,1: 378 (1910).

Young stems short, rusty-tomentose, glabrate with age. *Leaves* opposite-suboppo-site, lanceolate to ovate-elliptic, 35–50 × 15–20 mm, short rusty-pubescent along veins, otherwise whitish pubescent, penninerved, veins somewhat raised below; petioles 5–7 mm long, rusty-pubescent. *Inflorescence:* umbels solitary, axillary, 3–5-flowered, dens-ely rusty-tomentose; penducles 1–2 mm long, slightly shorter than pedicels. *Corolla* with conspicuous basal swelling and erect lobes, 40–45 mm long, tube split 10–12 mm below lobes, covered by a dense, conspicuous rusty-orange tomentum. *Filaments* with small tooth below anther. *Style* constricted below stigma. *Berries* unknown. *Flowering* Decem-ber through February. Fig. 6.

Parasitic on species of *Combretum, Parinari,* and *Terminalia* in northern South West Africa and eastern Caprivi.

Vouchers: *De Winter* 4268; *Killick & Leistner* 3193.

Balle in F.T.W.A. ed. 2, 1: 662 (1958) included *T. terminaliae* in the synonomy of *T. heteromorphus* (A.Rich.) Danser but in F.S.W.A. 22: 11 (1969) she kept the former name. This will have to be re-evaluated once the tropical African species become better known.

13. Tapinanthus natalitius (*Meisn.*) *Danser* in Verh. K. Akad. Wet., sect. 2, 29, 6: 116 (1933). Type: Natal, sine loc. exact., *Krauss* 208 (K, holo.!).

Loranthus natalitius Meisn. in Hooker, Lond. J. Bot. 2: 539 (1843); Harv. in F.C. 2: 576 (1862); Sprague in F.C. 5, 2: 114 (1915).

Relatively large shrubs 1 m or higher, glabrous to variously pubescent. *Older stems* glabrate. *Leaves* opposite, borne mostly on short spur-branches in 2–4 pairs, deciduous in winter (at least subsp. *zeyheri*), oblanceo-late to obovate-elliptic, (20–) 30–40 (–70) × 10–20 mm, lower leaves mostly shorter and broader than upper, glabrous, puberulent to shortly whitish pubescent, especially near petiole, when glabrous often glaucous; petioles subsessile, up to 3 mm long, often indistinct from the cuneate base. *Inflores-cence:* spur-branches 20–50 mm long, glab-rous to puberulent or shortly densely white-hirsute, bearing (2–) 3(–5) terminal flowers; pedicels 4–8 mm long, with same pubescence

type as spur-branches. *Corolla* with conspi-cuously swollen base 45–65 mm long, tube split 20–30 mm, mostly white, lobes yellow-red, glabrous to puberulent or pilose; mature buds with clavate apex shortly unci-nate. *Filaments* without tooth below anther. *Style* constricted below stigma. *Berries* orbicular, 12–15 mm long, dark red. *Flowering* October through February.

Parasitic primarily on species of *Acacia,* also on *Combretum* spp. Found from central and eastern Transvaal south to central Natal.

The species is divided into 2 subspecies, the typical form from Natal and subsp. *zeyheri* from central and eastern Transvaal. The pubescence of pedicels and leaves is a character too variable for effective separation of these taxa at specific level. As there appears to be some geographical restriction of these characters subspecific status seems appropriate.

Leaves mostly obovate, 40–50 mm long, glabrous and glaucous; corolla glabrous; pedicels glabrous to lightly canescent; primarily in Natal . . . .(a) subsp. *natalitius*

Leaves mostly oblanceolate-obovate, 20–30 mm long, mostly puberulent to pubescent; corolla puberulent to hirsute; pedicels puberulent to pubescent; primarily in Transvaal, Swaziland and northern Natal . . . .(b) subsp. *zeyheri*

(a) subsp. **natalitius.**

Loranthus natalitius Meisn. in Hooker, Lond. J. Bot. 2: 539 (1843); Harv. in F.C. 2: 576 (1862); Sprague in F.C. 5, 2: 114 (1915); Ross, Fl. Natal 152 (1972); Gibson, Wild Flow. Natal, pl. 30, 1 (1974). *Acranthemum natalitius* (Meisn.) v. Tieghem in Bull. Soc. bot. Fr. 42: 255 (1895).

Distinguished from subsp. *zeyheri* by the characters mentioned in the key. Fig. 6; Map 4.

Vouchers: *Acocks* 10751; *Sidey* 3548; *Stayner* 8958.

(b) subsp. **zeyheri** (*Harv.*) *Wiens* in Bothalia 12: 423 (1978). Type: Transvaal, Magaliesberg, *Zeyher* 751 (K, holo.!; S!; SAM!).

Loranthus zeyheri Harv. in F.C. 2: 576 (1862), pro parte excl. var. *minor*; Sprague in F.C. 5, 2: 113 (1915); Burtt Davy, Fl. Transv. 465 (1932); Letty, Wild Flow. Transv., pl. 61, 4 (1962). *Acranthemum zeyheri* (Harv.) v. Tieghem in Bull. Soc. bot. Fr. 42: 255 (1895). *Tapinanthus zeyheri* (Harv.) Danser in Verh. K. Akad. Wet., sect. 2, 29, 6: 122 (1933).

L. zeyheri Harv. var. *minor* Harv. in F.C. 2: 576 (1862). *Tapinanthus minor* (Harv.) Danser in Verh. K. Akad. Wet., sect. 2, 29, 6: 116 (1933). Type: Trans-vaal, Magaliesberg, *Zeyher* 571a (S, holo.!).

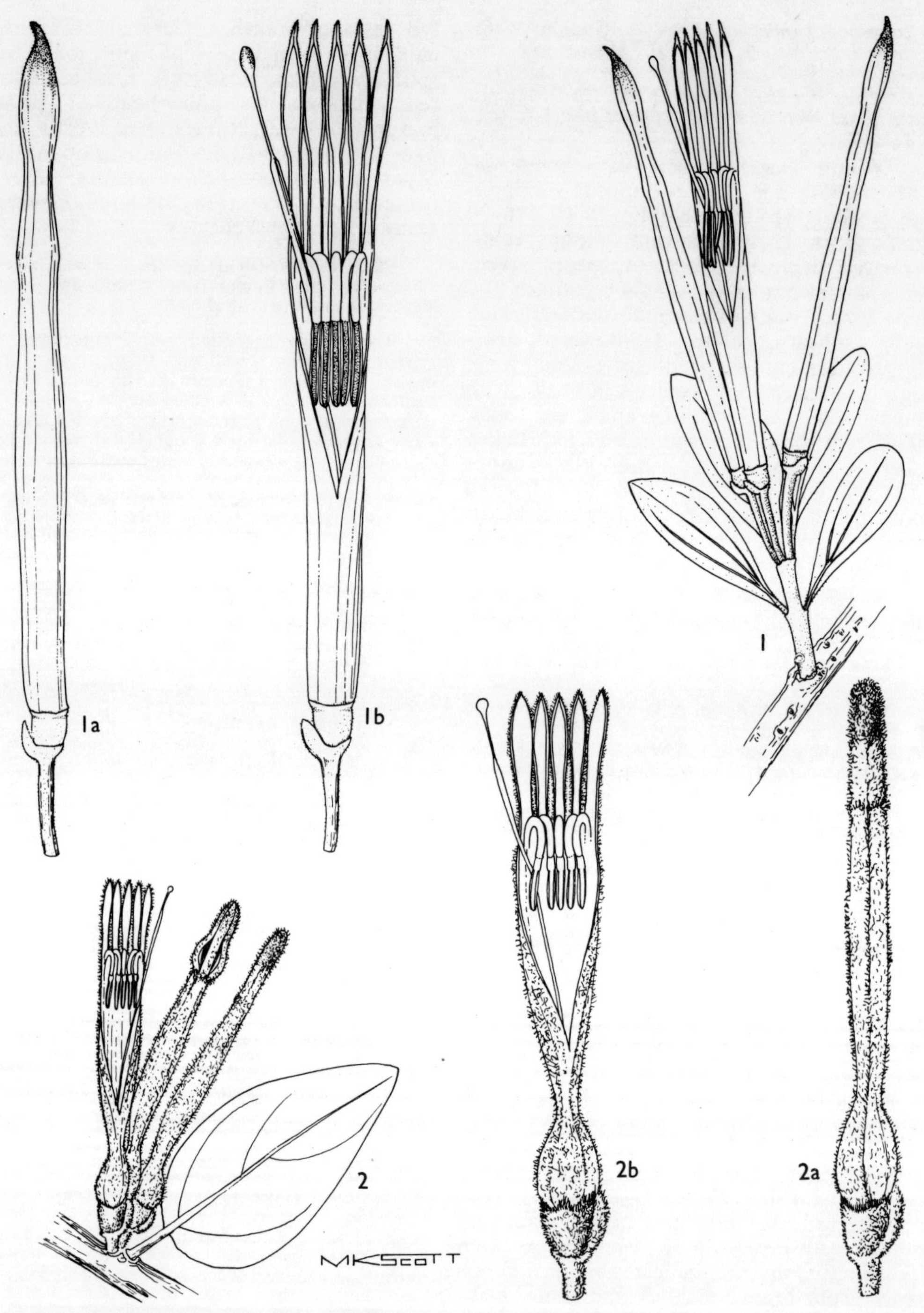

FIG. 6.—1, **Tapinanthus natalitius** subsp. **natalitius**, flowering twig, ×1; 1a, mature bud, ×1,5; 1b, flower, ×1,5 (*Green* G60A). 2, **T. terminaliae**, flowering twig, ×1; 2a, mature bud, ×1,5; 2b, flower, ×1,5 (*Vorster* 2786).

L. moorei Sprague in Kew Bull. 1915: 70, 80 (1915); in F.C. 5, 2: 114 (1915); Burtt Davy, Fl. Transv. 465 (1932). *Tapinanthus moorei* (Sprague) Danser in Verh. K. Akad. Wet., sect. 2, 29, 6: 116 (1933). Type: Transvaal, near Barberton, *Moore* s.n. (K, holo.!).

Differing from the typical subspecies primarily in the characteristics mentioned in the key (Map 4).

Vouchers: *De Winter* 410; *Galpin* 987; *Kinges* 1309.

Tapinanthus natalitius subsp. *zeyheri* is deciduous. This condition characterizes both this subspecies and *T. rubromarginatus,* with which it is sympatric. Sprague separated *T. moorei* from subspecies *zeyheri* on the basis of much enlarged bracts and glabrous pedicels, but these characters are not consistently correlated and appear to occur sporadically throughout the range of subspecies *zeyheri* and are thus of little taxonomic value. A form with consistently enlarged bracts and a distinctive pinkish, pilose corolla appears to occur in the Loskopdam Nature Reserve near Witbank.

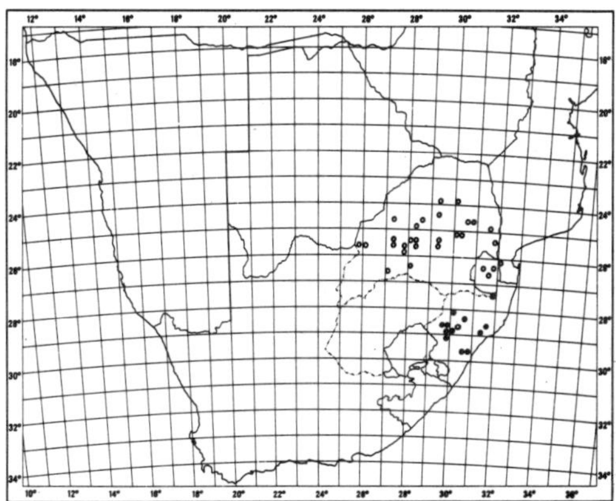

MAP 4.—●*Tapinanthus natalitius* subsp. **natalitius** ○T. natalitius subsp. **zeyheri**

14. Tapinanthus gracilis *Toelken & Wiens* in Jl S. Afr. Bot. 45:224 (1979). Type: Natal, 12 km N of Josini, *Tölken & Germishuizen* 5797 (PRE, holo.!).

Loranthus natalitius Meisn. var. *minor* (Harv.) Wood, Handb. Natal Pl. 115 (1907), pro parte typo excl. *L. minor* (Harv.) Sprague in Kew Bull. 1915: 80 (1915), pro parte; in F.C. 5, 2: 115 (1915), pro parte quoad spec. enum.

Shrubs, mostly less than 1 m high, essentially glabrous. *Stems* greyish brown. *Leaves* opposite-subopposite, mostly lanceolate to ovate-elliptic, 15–30 × 5–15 mm, relatively chartaceous, often with 3 distinct basal veins, borne mostly on short spur-branches 10–30 mm long, in 2–4 pairs, basal pair often shorter and broader than upper; petioles subsessile to 4 mm long, often indistinct into the cuneate base. *Inflorescence:* flowers originating terminally from spur-branches in groups of 2–6; pedicels 6–8 mm long. *Corolla* without conspicuously swollen base, dark red with yellow band below apex, 35–40 mm long, tube split 7–9 mm below erect lobes; mature buds with apex shortly uncinate. *Berries* obovoid, 8–10 mm long, red. *Flowering* mostly from November through February. Fig. 7.

Parasitic on a large number of diverse hosts including species of *Acacia, Acalypha, Dombeya, Ehretia, Zanthoxylum, Grewia, Maytenus, Olea, Berchemia, Plumbago, Tarchonanthus* and *Viscum*; occurring from the eastern Transvaal and Swaziland to southern Natal (Map 5).

Vouchers: *Rudatis* 765; 1120; *Stauffer & Weder* 5277; *Strey* 2074; 4445; 9567.

Although apparently parasitic on numerous host genera, in northern Natal this species was collected only on another mistletoe, *Viscum verrucosum.* Epiparasitism of this type is known in a number of mistletoes (see discussion under *T. ceciliae* and *T. kraussianus*). Surprisingly, the phenomenon is not always simple to detect and collectors unfamiliar with such mistletoes may easily overlook the actual point of attachment and mistakenly record the tree as the primary host. In view of the large number of trees and shrubs reported as hosts for *T. gracilis*, a careful analysis of host-parasite ecology in this species is needed.

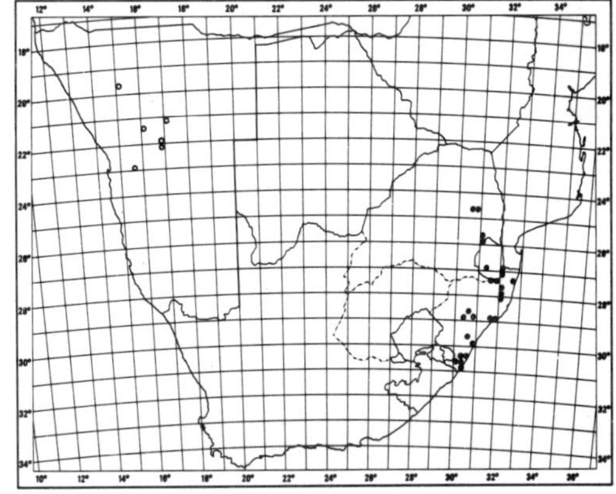

MAP 5.—●**Tapinanthus gracilis** ○T. **discolor**

15. Tapinanthus discolor (*Schinz*) Danser in Verh. K. Akad. Wet., sect. 2, 29, 6: 110 (1933). Syntypes: South West Africa, Rehoboth, *Fleck* 452 (Z!); *Fleck* 881 (Z).

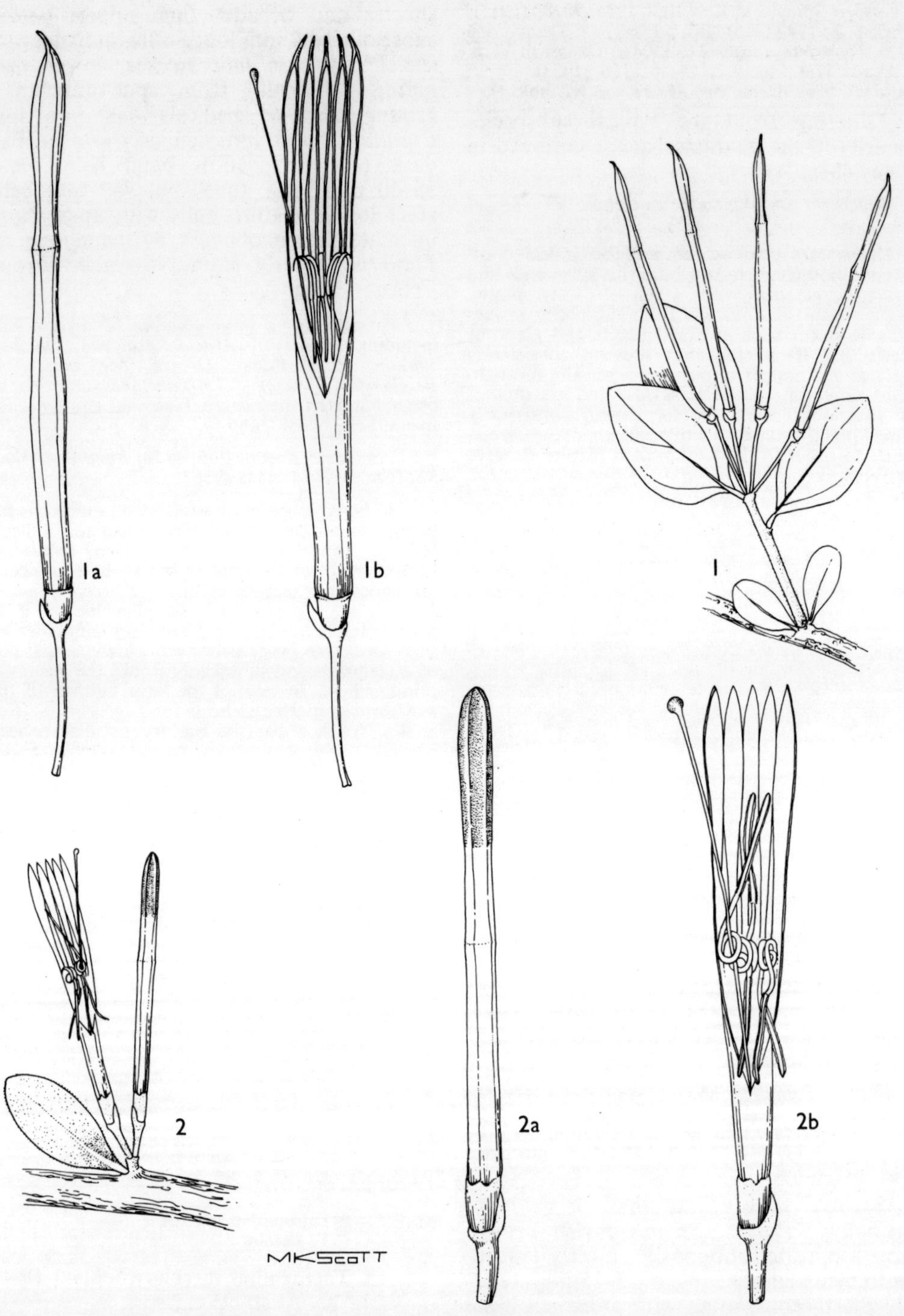

FIG. 7.—1, **Tapinanthus gracilis**, flowering twig, ×1; 1a, mature bud, ×2; 1b, flower, ×2 (*Ward* 1905). 2, **T. discolor**, flowering twig, ×1; 2a, mature bud, ×2; 2b, flower, ×2 (*Jensen* 413).

Loranthus discolor Schinz in Bull. Herb. Boissier sér. 1, 4, App. 3: 52 (1896). *Agelanthus discolor* (Schinz) Balle in Mitt. bot. StSamml., Münch. 7: 157 (1968); in F.S.W.A. 22: 3 (1969).

L. bosciae Engl. & Krause in Bot. Jb. 43: 401 (1909); Sprague in F.T.A. 6, 1: 317 (1910). Type: South West Africa, Okahandja, *Dinter* 284 (SAM!).

L. juttae Dinter, Fl. Deutsch-Südw. Afr. 56 (1909). Type: unknown.

Shrubs, perhaps 0,5–1 m high. *Older stems* relatively thick, greyish brown with greatly swollen nodes, young branches minutely puberulous. *Leaves* alternate, oblanceolate-oblong, 15–25 × 4–7 mm, minutely puberulent to glabrous, venation inconspicuous to distinctly 3-nerved; petioles 2 mm long to subsessile. *Inflorescence:* umbels axillary, sessile, 2–4-flowered; pedicels 1–2 mm long, puberulent. *Corolla* without conspicuous basal swelling, greenish to pink, 38–42 mm long, tube split 10–12 mm below lobes; lobes erect, crimson, approximately as long as rest of corolla; calices tubular, 2 mm long. *Filaments* without small tooth below anthers. *Style* filiform. *Berries* unknown *Flowering* in October through December Fig. 7.

Parasitic on species of *Boscia* in central and northern South West Africa (Map 5).

Vouchers: *Niehaus* 13496; *Volk* 2629; *Wiss* 731.

16. **Tapinanthus lugardii** (*N.E.Br.*) *Danser* in Verh. K. Akad. Wet., sect. 2, 29, 6: 115 (1933). Syntypes: Botswana, Kwebe Hills, *Mrs Lugard* 20 (K!); Totin, *Lugard* 32 (K!).

Loranthus lugardii N.E.Br. in Kew Bull. 1909: 135 (1909), as *lugardi*; Sprague in F.T.A. 6, 1: 318 (1910), as *lugardi*.

L. breyeri Bremek. in Ann. Transv. Mus. 15: 238 (1933). Type: Transvaal, near Pietersburg, *Bremekamp & Schweickerdt* 33 (PRE, holo.!).

Shrubs up to perhaps 1 m high. *Young stems* densely puberulent, glabrate and greyish with age. *Leaves* mostly alternate, occasionally in fascicles, mostly oblanceolate, 20–35 (–60) × 5–10 (–15) mm, puberulous to glabrate, coriaceous, sessile-subsessile from the cuneate base. *Inflorescence:* flowers sessile to subsessile, 1–4 in axils; corolla without conspicuous basal swelling, pale greenish yellow to pink with age, 33–37 mm long, tube split 8–10 mm below filaments; lobes essentially erect, but recurved about 45 degrees or less; calyx tubular, 2 mm long. *Filaments* without tooth below anther. *Style* filiform. *Berries* ellipsoid, red-orange, 8–10

mm long, persistent calyx conspicuous. *Flowering* from November through January. Fig. 8.

Parasitic primarily on *Acacia* spp., but also found on species of *Nicotiana* and *Ximenia*; occurring from northern Botswana to central Transvaal and northern Natal (Map 6).

Vouchers: *Barnard* 563A; 580A; *Mogg* 24506; *Ward* 3868.

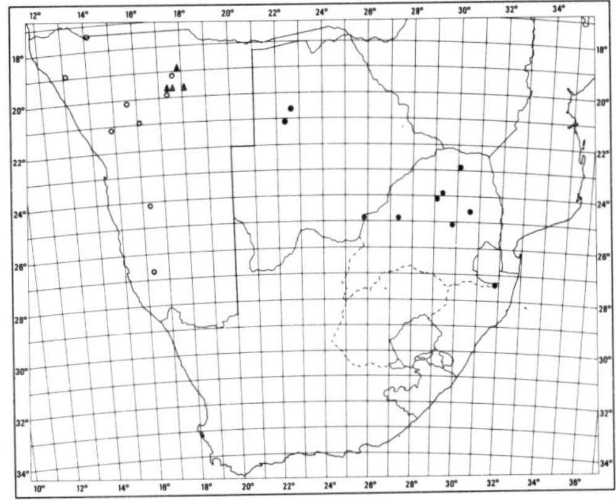

MAP 6.—●**Tapinanthus lugardii**
○T. **guerichii**
▲T. **cinereus**

17. **Tapinanthus guerichii** (*Engl.*) *Danser* in Verh. K. Akad. Wet., sect. 2, 29, 6: 113 (1933). Type: South West Africa, Karibib, *Gürich* 35 (B, holo.; PRE, photo.!).

Loranthus guerichii Engl. in Bot. Jb. 19: 130 (1894); Sprague in F.T.A. 6, 1: 297 (1910). *Phragmanthera guerichii* (Engl.) Balle in Mitt. bot. StSamml., Münch. 7: 169 (1968); in F.S.W.A. 22: 7 (1969).

Stems relatively thick, reddish brown, conspicuously furrowed (at least when dry), canescent when young, glabrate with age. *Leaves* fascicled on the swollen nodes, mostly elliptic-oblong to oblanceolate, 10–20 × 4–8 mm, densely stellate-pubescent on both blade and petiole; petiole nearly half as long as blade, 5–7 mm long. *Inflorescence:* flowers arising mostly in pairs from leaf fascicles; pedicels 2 mm long (mostly obscured by the dense tomentum); bracts long linear, 5–10 mm long, at least twice or more the length of ovary and calyx, both leaves and associated flowers densely whitish tomentose. *Corolla* without conspicuous basal swelling, 55–60 mm long, tube split 20–22 mm below erect lobes. *Filaments* without

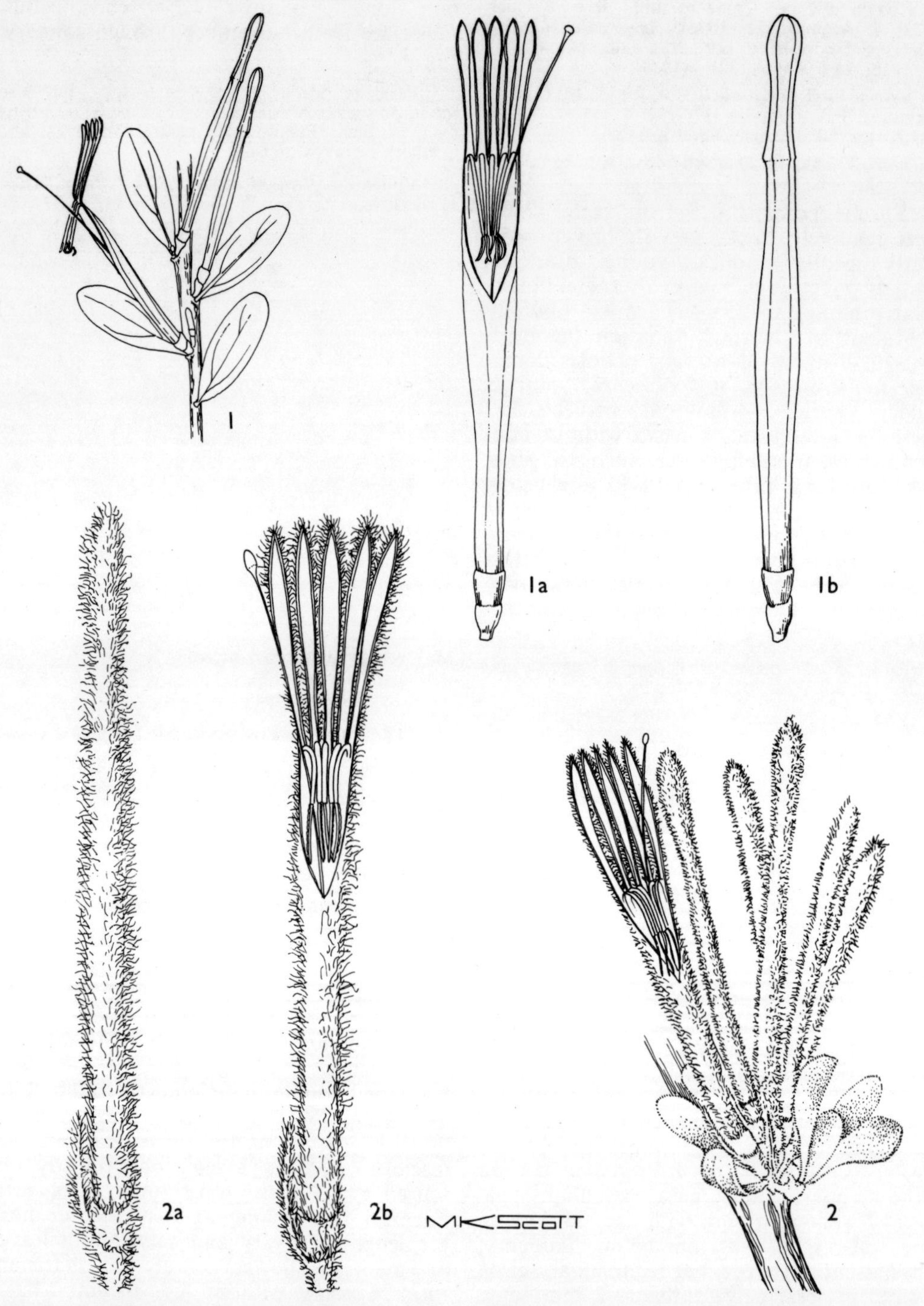

FIG. 8.—1, **Tapinanthus lugardii,** flowering twig, ×1; 1a, mature bud, ×2; 1b, flower, ×2 (*Louw* 811). 2, **T. guerichii,** flowering twig, ×1; 2a, mature bud, ×1,5; 2b, flower, ×1,5 (*Obermeyer* in TRV 32813).

tooth below anthers. *Style* somewhat constricted below stigma. *Berries* moderately stellate-pubescent, ellipsoid, 10–12 mm long. *Flowering* from September through December. Fig. 8.

Parasitic on species of *Commiphora* and *Euphorbia* in northern, central and southern South West Africa (Map 6).

Vouchers: *Abner* 29; *Campbell & Sacks* 5000; *Giess, Volk & Bleissner* 5081.

18. **Tapinanthus cinereus** (*Engl.*) *Danser* in Verh. K. Akad. Wet., sect. 2, 29, 6: 111 (1933). Type: Angola, near S. Joao do Croque, Rio Croque, Cabo Negro, *Welwitsch* 4884 (PRE!).

Loranthus cinereus Engl. in Bot. Jb. 20: 103 (1894); Hiern in Cat. Afr. Pl. Welw. 1, 4: 928 (1900); Sprague in F.T.A. 6, 1: 296 (1910). *Phragmanthera cinerea* (Engl.) v. Tieghem ex Durand & B. D. Jackson, Kew Ind., suppl. 1: 326 (1906); Balle in Mitt. bot. StSamml., Münch. 7: 165 (1968); in F.S.W.A. 22: 6 (1969).

L. fulvus Engl. in Bot. Jb. 20: 103 (1894), non Korth. (1839). *Phragmanthera fulva* (Engl.) v. Tieghem ex Durand & B. D. Jackson, Kew Ind., suppl. 1: 326 (1906).

L. dombeyae Krause & Dinter in Bot. Jb. 45: 283 (1910). *Tapinanthus dombeyae* (Krause & Dinter) Danser in Verh. K. Akad. Wet., sect. 2, 29, 6: 111 (1933). Type: South West Africa, Otavi, *Dinter* 933 (SAM!).

Young branches densely white-tomentose, glabrate with age. *Leaves* opposite-subopposite on young lateral shoots, or fascicled on swollen nodes of older stems, mostly ovate-oblong, 20–30 × 15–20 mm, densely white-tomentose (as young branches) when immature, lightly scattered, short, stellate-pubescent when mature, especially midrib and petiole; petioles 4–6 mm long. *Inflorescence:* umbels mostly axillary, 2–4-flowered; peduncles 1–2 mm long, approximately equalling the pedicels; mature buds with clavate apex winged along sutures. *Corolla* without conspicuous basal swelling, short, floccose, stellate-pubescent at anthesis, yellow-orange, tube split 9–11 mm below lobes, 35–44 mm long; lobes erect, spathulate. *Filaments* lacking tooth at base of **filament**. *Berries* unknown. *Flowering* in **December** and February (and probably also **other** times). Fig. 9.

Hosts unknown; occurring in northern South West Africa to Angola (Map 6).

Vouchers: *Dinter* 5239; *Giess & Smook* 10613; *Welwitsch* 4884.

The combinations *Phragmanthera cinerea* and *P. fulva* were intended but not validly published by v· Tieghem (1895).

19. **Tapinanthus glaucocarpus** (*Peyr.*) *Danser* in Verh. K. Akad. Wet., sect. 2, 29, 6: 112 (1933). Type: Angola, Benguella, *Wawra* 287 (W, holo.).

Loranthus glaucocarpus Peyr. in Sber. Akad. Wiss. Wien, Naturwiss. Kl. 38: 571 (1860); Sprague in F.T.A. 6, 1: 295 (1910); 1028 (1913). *Phragmanthera glaucocarpa* (Peyr.) Balle in Mitt. bot. StSamml., Münch. 7: 167 (1968); in F.S.W.A. 22: 7 (1969).

L. cistoides Welw. ex Engl. in Bot. Jb. 20: 103 (1894). *Phragmanthera cistoides* (Welw. ex Engl.) v. Tieghem ex Durand & B. D. Jackson, Kew Ind., suppl. 1: 326 (1906). *Tapinanthus cistoides* (Welw. ex Engl.) Danser in Verh. K. Akad. Wet., sect. 2, 29, 6: 110 (1933). Syntypes: Angola, Pungo Andongo, *Welwitsch* 4847 (K!); near Benguella, *Welwitsch* 4853 (K!); near Cazimba, *Welwitsch* 4857 (K!); Pungo Andongo, *Teuscz* sub v. Mechow 90.

L. cistoides Welw. ex Engl. var. *longiflora* Schinz in Bull. Herb. Boissier sér. 1, 4, App. 3: 52 (1896). Syntypes: South West Africa, Grootfontein, *Schinz* 294 (Z!); sine loc. exact., *Höpfner* 123 (Z!).

L. otaviensis Engl. & Krause in Bot. Jb. 45: 285, fig. 1 (1910), as *otavensis*. Type: South West Africa, Otavi, *Dinter* 901 (K!; SAM!).

Young stems densely long tomentose, glabrate with age. *Leaves* opposite-subopposite on young shoots, or fascicled on nodes of older stems, ovate-oblong, 25–35 × 15–20 mm, young leaves densely long tomentose (as young stems), shortly densely stellate to scattered stellate at maturity; petioles 1–2 mm long to subsessile, densely tomentose. *Inflorescence:* umbels axillary or associated with leafy fascicles, 2–6-flowered; peduncles 2–3 mm long, about twice as long as pedicels. *Corolla* without conspicuous basal swelling, whitish from long, spreading hairs (2–3 mm long) covering entire corolla, 45–50 mm long, tube split 10–12 mm below lobes; lobes erect, spathulate, with membranous margins; bracts becoming foliaceous, up to 10 mm long. *Filaments* without tooth below anthers. *Style* essentially filiform. *Berries* ovoid, c. 15 mm long. *Flowering* in May, August, December and no doubt also at other times. Fig. 9.

Parasitic on *Croton* spp. (and probably also other genera) in northern and central South West Africa and in Angola (Map 7).

Vouchers: *Dinter* 5240; *Giess* 10162; *Volk* 2750.

V. Tieghem (1895) referred *Loranthus cistoides* to the genus *Phragmanthera* but did not actually make the combination.

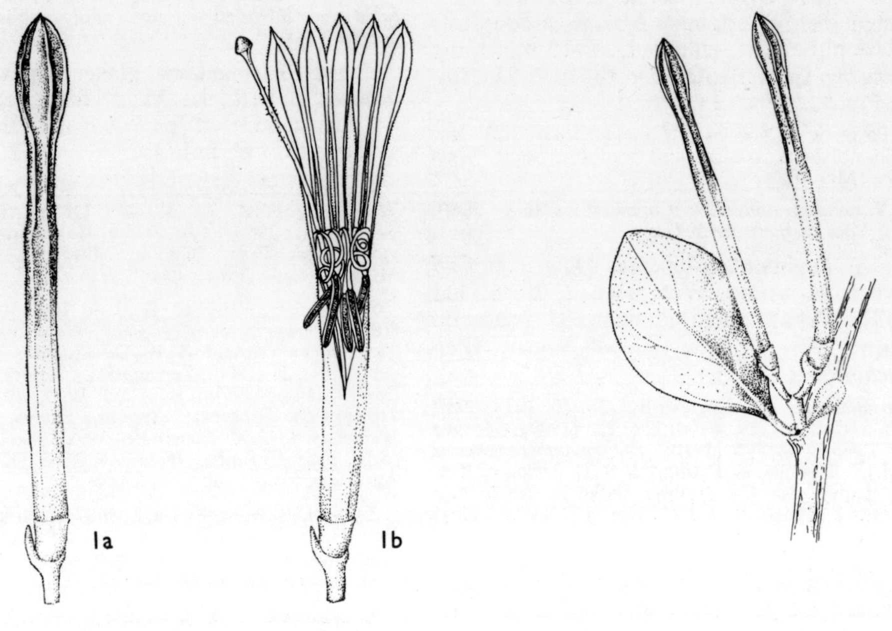

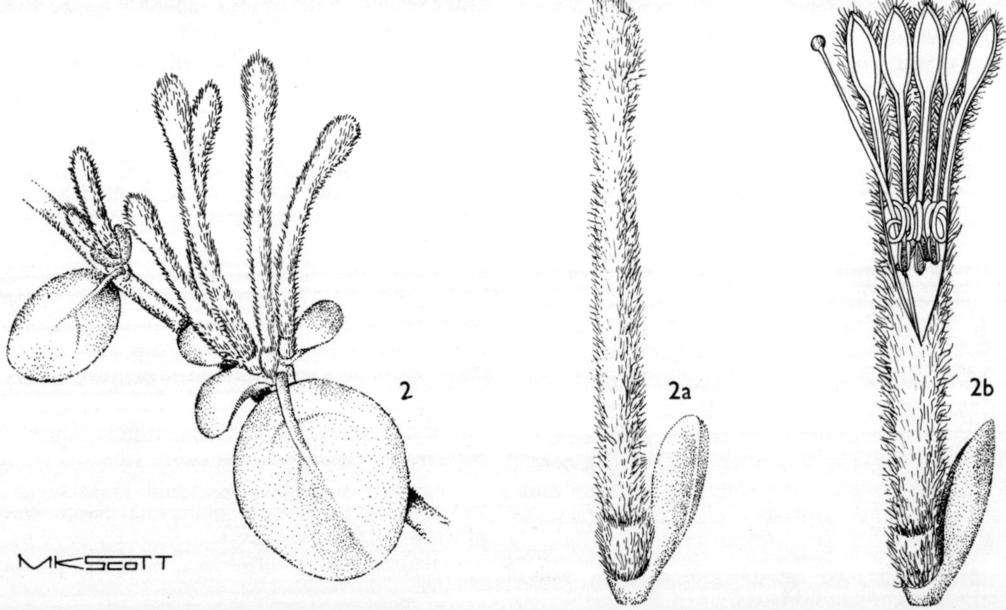

FIG. 9.—1, **Tapinanthus cinereus**, flowering twig, ×1; 1a, mature bud, ×1,5; 1b, flower, ×1,5 (*Schoenfelder* 291). 2, **T. glaucocarpus**, flowering twig, ×1; 2a, mature bud, ×1,5; 2b, flower, ×1,5 (*Dinter* 5240).

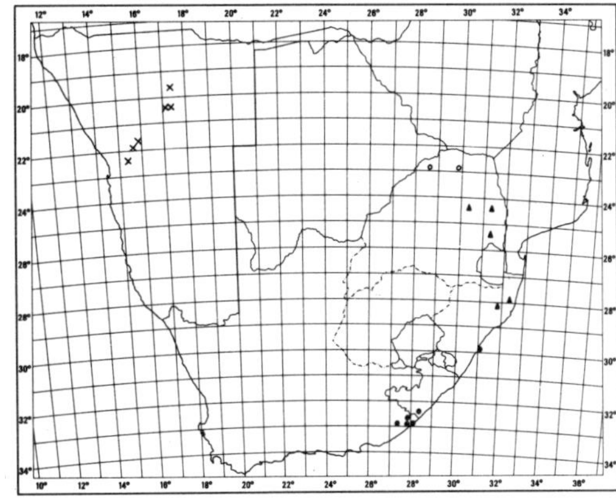

2074b   2. **TIEGHEMIA**

Tieghemia *Balle* in Bull. Séanc. Inst. r. colon. belge, n.s. 2: 1062 (1956). Type species:
T. quinquenervius (Hochst.) Balle.

Loranthus sect. *Dendrophthoe* (Mart.) Engl. in Pflanzenfam. ed. 1, 3, 1: 186 (1894), pro parte; Engl. & Krause
in Pflanzenfam. ed. 2, 16b: 162 (1933), pro parte quoad 'Gruppe' *Quinquenerves*.

L. sect. *Quinquenerves* Sprague in Kew Bull. 1915: 70 (1915).

Shrubs of relatively small size, usually less than 0,5 m high, glabrous. *Stems* buff-coloured
with usually dense lenticels and often greatly swollen, floriferous nodes; wood a dull pink
(in living plants). *Haustoria* expanding laterally and proximally along cambium from point
of infection and enlarging with age and penetrating into xylem and erupting through cortex
near shoots. *Leaves* alternate, blades mostly elliptic to suborbicular. *Inflorescence* mostly a
3–6-flowered umbel with short peduncles and pedicels less than 2 mm long, solitary in axils
or fascicled on older swollen nodes. *Flowers* 5-merous, gamopetalous, bilaterally symmetrical
by the presence of a "V"-shaped, unilateral split approximately as long as erect lobes. *Corolla*
with conspicuous basal swelling and bands of red, white, pink, and pale green in various
combinations. *Filaments* coiled or curved at anthesis, with 2 minute (0,5 mm) lateral appen-
dages or teeth present 1–2 mm below anthers. *Style* filiform; stigma capitate. *Flowering*
mostly during winter (April–August).

A genus of 2 or 3 species mostly in Southern Africa, but also in adjoining Mozambique and probably
Rhodesia.

Leaves dark green, elliptical to oblong-rounded, mostly 35–50×15–25 mm, mostly 3–5-nerved from just
    above base; petioles 4–6 mm long, indistinctly demarcated from cuneate base, often winged:

    Calyx rim extending 1 mm or more above highest point of bracts; flowers mostly associated with leafy
        shoots ....1. *T. quinquenervius*

    Calyx rim only slightly, if at all, higher than bracts; flowers often fascicled on swollen nodes of older
        branches in addition to occurring in axils of leafy shoots ....2. *T. bolusii*

Leaves light greyish-green, lanceolate-ovate, mostly 18–25×8–12 mm, venation inconspicuous except
    occasionally for the midrib; petioles 2–3 mm long, clearly distinct from the usually rounded base,
    never winged ....3. *T. rogersii*

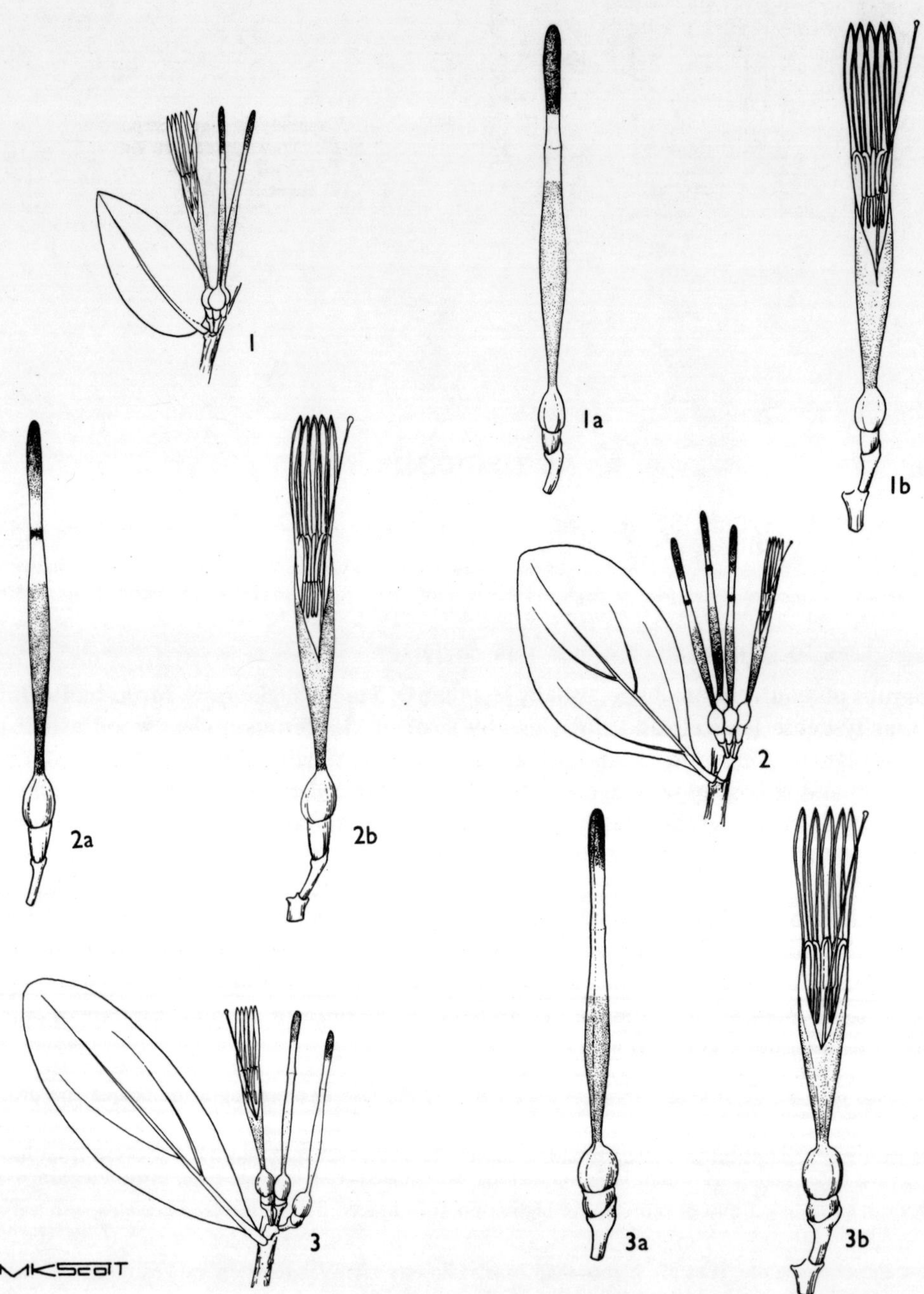

FIG. 10.—1, **Tieghemia rogersii**, flowering twig, ×1; 1a, mature bud, ×2; 1b, flower, ×2 (*Wiens* 5317). 2, **T. quinquenervius**, flowering twig, ×1; 2a, mature bud, ×2; 2b, flower, ×2 (*Wiens* 5370). 3, **T. bolusii**, flowering twig, ×1; 3a, mature bud, ×2; 3b, flower, ×2 (*Wiens* 5426).

1. Tieghemia quinquenervius (*Hochst.*)
Balle in Bull. Séanc. Inst. r. colon. belge, n.s.
2: 1065 (1956), as *quinquenervis*. Type: Natal,
Port Natal, *Krauss* s.n. (not traced).

Loranthus quinquenervius Hochst. in Flora 27: 432
(1844); Harv. in F.C. 2: 578 (1862); Wood & Evans,
Natal Plants pl. 295 (1902); Sprague in F.C. 5, 2:
111 (1915), as *quinquenervis*; Burtt Davy, Fl. Transv.
465 (1932), do; Batten & Bokelmann, Wild Flow. E.
Cape Prov. pl. 55, 6 (1966), do; Ross in Fl. Natal 152
(1972), do; Gibson, Wild Flow. Natal pl. 30, 3 (1975),
do. *Tapinanthus quinquenervius* (Hochst.) Danser in
·Verh. K. Akad. sect. 2, 29, 6: 118 (1933).

Plants glabrous. *Leaves* alternate, mostly
elliptical to ovate-suborbicular, 40–50 × 15–25
mm, usually 5-nerved from just above the
base; petioles 3–5 mm long, cuneate into the
base, usually slightly winged. *Inflorescence:*
umbels mostly solitary in axils, 4–6-flowered;
peduncles less than 1 mm long; pedicels
1–2 mm long. *Corolla* with dark red and
white bands, 35–40 mm long: apical sixth red,
next two sixths white [but interrupted by a
narrow (2 mm) red band at point of filament
attachment], next 2 sixths red, and basal
sixth white (i.e., the basal swelling), tube split
10–12 mm below point of filament attachment.
Berries ellipsoid, 8–10 mm long, bright red.
Flowering from approximately (April) June–
August (September); $n=9$. Fig. 10.

Parasitic on species of *Cassine, Celtis,* and probably other hosts; from eastern Cape to central coastal
Natal (Map 7).

Vouchers: *Acocks* 21674; *Galpin* 1832; *Pegler*
1517.

Previous workers have not recognized that the
corolla of this species possesses a unilateral split and
is thus zygomorphic. This characteristic is easily
distorted in drying if the buds are mature but not yet
actually open. Under these circumstances the buds
often open on drying, but the split does not become
evident and the corolla appears regular.

2. Tieghemia bolusii (*Sprague*) *Wiens* in
Bothalia 12: 423 (1978). Type: Mozambique,
Delagoa Bay, *Bolus* 9764 (K, holo.!; PRE!).

Loranthus bolusii Sprague in Kew Bull. 1915: 81
(1915); Ross, Fl. Natal 152 (1972). *Tapinanthus bolusii*
(Sprague) Danser in Verh. K. Akad. Wet., sect. 2,
29, 6: 108 (1933).

Closely related to *T. quinquenervius* and
possibly best considered as conspecific with it.
The taxon differs from *T. quinquenervius* by
its shorter and broader calyx tube (see key)
and both the colour and pattern of its corolla
markings. In *T. bolusii* the red is bright instead
of dark and a very light green is substituted
for the white. Flowering from approximately
April to November. Fig. 10.

Parasitic on species of *Cassine, Berchemia,* and
undoubtedly other genera from coastal northern
Natal to the north-eastern Transvaal and Mozambique
(Map 7).

Voucher: *Wiens & Van Wyk* 5426.

Field work should be completed in central Natal
to determine if this taxon is sympatric with *T. quinquenervius,* and if so, whether they intergrade.

3. Tieghemia rogersii (*Sprague ex Burtt
Davy*) *Wiens* in Bothalia 12: 423 (1978).
Type: Transvaal, Soutpansberg, Waterpoort,
Rogers 21507 (K, holo.!; PRE!).

Loranthus rogersii Sprague ex Burtt Davy, Fl.
Transv. 465 (1932).

Closely related, but quite distinct from
T. bolusii by the smaller, light grey-green,
veinless leaves, with a short, but distinct, non-
winged petiole (see key). The corolla banding
also differs from *T. bolusii.* Flowering from
April to August; $n=9$. Fig. 10.

Vouchers: *Van der Schijff* 6622; *Wiens* 5317.

The species is known only from the north-central
Transvaal; probably also in Rhodesia and Mozambique (Map 7).

Burtt Davy attributes this species to Sprague, but
there is no evidence that Sprague actually published
the name. The type specimen at Kew, however, bears
Sprague's annotation. Burtt Davy's listing of dis-
tinguishing characters in his key and his designation
of a type specimen is sufficient documentation for him
to be considered the publishing author.

2074c   **3. MOQUINELLA**

Moquinella *Balle* in Bull. Séanc. Inst. r. colon., belge 25: 1628 (1954). Type species: *M. rubra*
(Spreng. f.) Balle.

Moquinia Spreng. f., Tent. Suppl. Syst. Veg. 9 (1828); Danser in Verh. K. Akad. Wet., sect. 2, 29, 6: 55 (1933)
(nom. rejic.) non DC. (1838). *Loranthus* sect. *Moquinia* (Spreng. f.) Sprague in Kew Bull. 1914: 367 (1914);
1915; 69 (1915).

Loranthus sect. *Dendrophthoe* (Mart.) Engl. in Pflanzenfam. ed. 1, 3, 1: 186 (1894), pro parte; in Bot. Jb. 20:
83 (1894), pro parte quoad 'Gruppe' *Oleaefolia*; Engl. & Krause in Pflanzenfam. ed. 2, 16b: 154 (1933), pro
parte quoad 'Gruppe' *Moquinia*. L. subgen. *Dendrophthoe* (Mart.) Engl. in Pflanzenfam. ed. 1, Nachtr. 1: 131
(1897), pro parte quoad 'Gruppe' *Lichtensteinia*.

Lichtensteinia sensu v. Tieghem in Bull. Soc. bot. Fr. 42: 254 (1895), non Wendl.

Shrubs perhaps 1 m high, essentially glabrous. *Leaves* alternate, subopposite, occasionally fascicled, somewhat coriaceous with age, mostly elliptic-linear to lanceolate (greatly variable in size), (15–) 30–40 (–80) × 5–10 mm; petioles 2–4 mm long. *Inflorescence* a 3–6-flowered umbel, or occasionally racemose by the addition of several flowers lower on the rachis, axillary or solitary on older stems. *Flowers* 5-merous; mature buds 30–40 mm long, orange basally, yellow above, the apical portion blackish, mostly cylindrical but slightly expanded basally. *Corolla* gamopetalous, bilaterally symmetrical (the tube bearing a short, "V"-shaped, unilateral split), lobes slightly longer than rest of corolla and spirally coiled at anthesis. *Filaments* attached near base of lobes, anthers normally breaking from filament at anthesis (the flower opens explosively). *Style* filiform; stigma capitate. Flowering from May through August; $n=9$.

A monotypic genus confined to the Cape Province.

Moquinella rubra (*Spreng. f.*) *Balle* in Bull. Séanc. Inst. r. colon. belge 25: 1630 (1954). Type: Cape, Uitenhage, *Zeyher* 296 (S,? holo.!; PRE, photo.!; see note below).

Moquinia rubra Spreng. f., Tent. Suppl. Syst. Veg. 9 (1828).

Loranthus elegans Cham. & Schlechtd. in Linnaea 3: 209 (1828). Sprague in F.C. 5, 2: 108 (1915); Batten & Bokelmann, Wild Flow. E. Cape Prov. pl. 55, 2 (1966). *Dendrophthoe elegans* (Cham. & Schlechtd.) Mart. in Flora 1: 109 (1830). *Scurrula elegans* (Cham. & Schlechtd.) G. Don, Gen. Syst. 3: 424 (1834). *Lichtensteinia elegans* (Cham. & Schlechtd.) v. Tieghem in Bull. Soc. bot. Fr. 42: 254 (1895). *Loranthus oleifolius* (Wendl.) Cham. & Schlechtd. var. *elegans* (Cham. & Schlechtd.) Harv. in F.C. 2: 577 (1862). Syntypes: Cape, *Mundt* s.n. (K!); *Krebs* s.n.

L. schlechtendalianus Schult., Syst. Veg. Mant. 7, 2: 1635 (1830), nom. superfl., in locum *L. elegantis* Cham. & Schlechtd. (1828) pro *L. eleganti* Mart. ex Schult. (1829).

L. glaucus Thunb. var. *burchellii* DC., Prodr. 4: 303 (1830). Type: Cape, Sundays River, *Burchell* 2887 (K!).

L. glaucus sensu DC., Prodr. 4: 303 (1830), non Thunb.

L. oleifolius sensu Eckl. & Zeyh., Enum. 358 (1834); sensu Presl, Bot. Bemerk. 76 (1844); sensu Marloth, Fl. S.A. 1, t. 38 A (1913), non Cham. & Schlechtd.

L. croceus E. Mey. ex Drège, Zwei Pfl. Doc. 63, 109, 139, 200 (1844), nom. nud.

The single species with characteristics of the genus. Fig. 11.

Parasitic on *Acacia, Euclea, Ficus, Grewia, Rhus, Diospyros* and *Salix* in the southern Cape region; disjunct in the north-western Cape (Map 8).

Vouchers: *Flanagan* 727; *Galpin* 1812; 10651; *Smith* 2797a.

A distinct monotypic genus, possibly representing an offshoot of *Tapinanthus* or perhaps an isolated evolutionary relic of uncertain ancestry. The nomenclature of this species has had a long and complex history; for reviews see particularly Sprague (Kew Bull. 1914: 359; 1914) and Danser (Verh. K. Akad. Wet., Sect. 2, 29, 6: 95; 1933).

It is uncertain whether the specimen in Stockholm herbarium is the type of this species as the collector's number cannot be found on it. It is, however, inscribed 'Moquinia rubra Spr., nov. gen.' and Stafleu (Regnum Veg. 52: 455; 1967) records Sprengel specimens of other families in Sonder's herbarium.

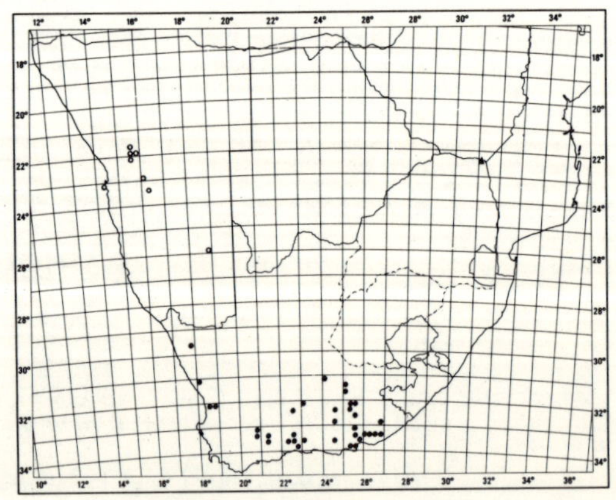

MAP 8.—● **Moquinella rubra**
○ **Odontella welwitschii**
▲ **Vanwykia remota**

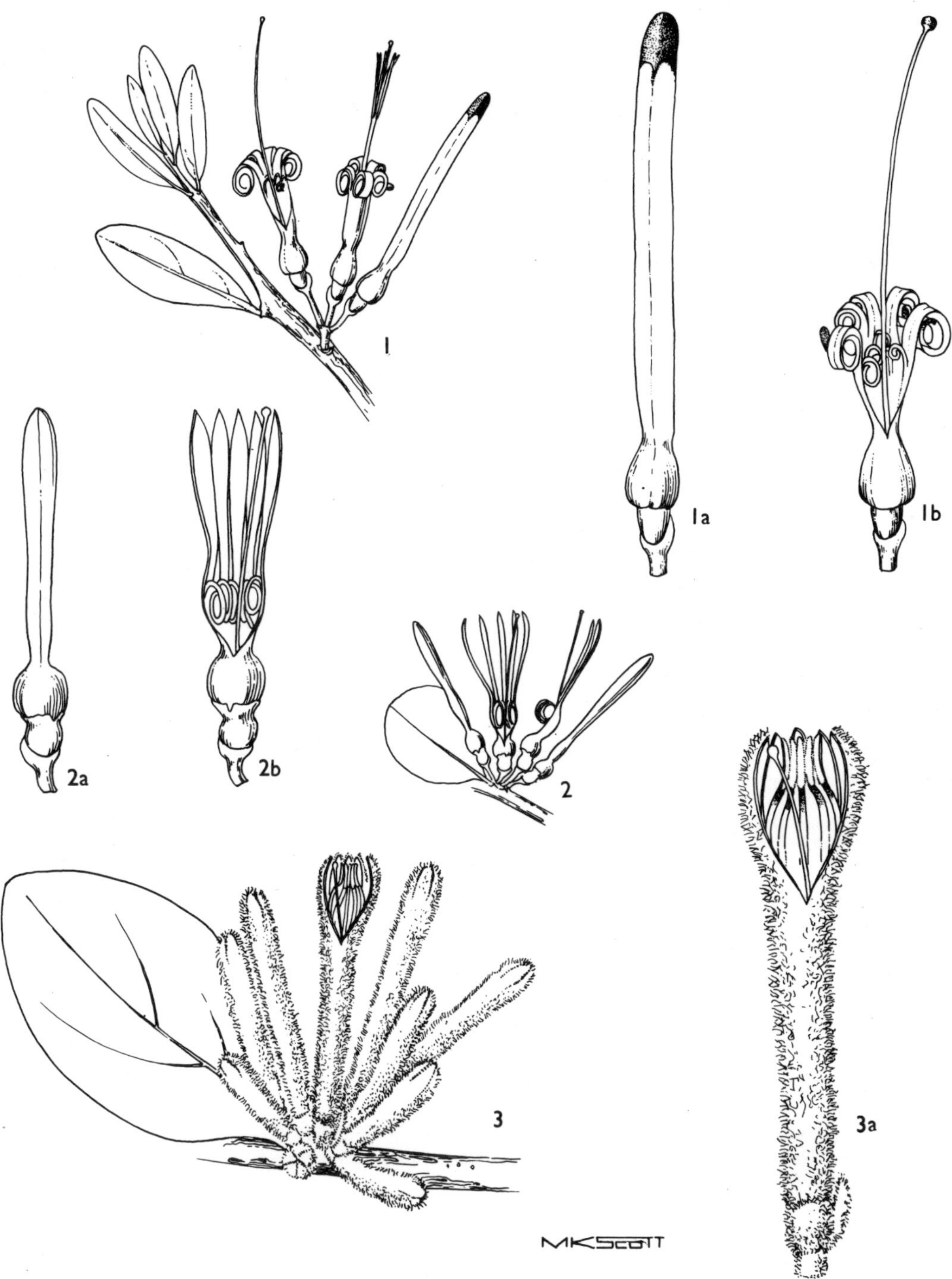

FIG. 11.—1, **Moquinella rubra**, flowering twig, ×1; 1a, mature bud, ×2; 1b, flower, ×2 (*Wiens* 5392). 2, **Odontella welwitschii**, flowering twig, ×1; 2a, mature bud, ×2; 2b, flower, ×2 (*Merxmüller & Giess* 30309). 3, **Vanwykia remota**, flowering twig, ×1; 3a, flower, ×2 (*Wiens* 5333).

2074d   4. **ODONTELLA**

Odontella *v. Tieghem* in Bull. Soc. bot. Fr. 42: 259 (1895); Balle in Bull. Séanc. Acad. r. Sci. colon. (outre Mer), sér. 2, 6: 1072 (1957), emend.; in F.S.W.A. 22: 4 (1968). Type species: *O. schimperi* (Hochst. ex A. Rich.) v. Tieghem.

Loranthus sect. *Dendrophthoe* (Mart.) Engl. in Pflanzenfam. ed. 1, 3, 1: 186 (1894), pro parte; in Bot. Jb. 20: 86 (1894), pro parte quoad 'Gruppe' *Rigidiflori*. *L.* sect. *Rigidiflori* Engl. ex Sprague in F.T.A. 6, 1: 266 (1910).

L. sect. *Tapinanthus* Blume, Fl. Jav. Loranth. 15 (1830); Engl. in Pflanzenfam. ed. 1, 3, 1: 187 (1894), pro parte; in Bot. Jb. 20: 110 (1894), pro parte quoad 'Gruppe' *Coriaceifolii*. *L.* sect. *Coriaceifolii* Engl. ex Sprague in F.T.A. 6,1: 266 (1910).

Oncocalyx v. Tieghem in Bull. Soc. bot. Fr. 42: 258 (1895), pro parte quoad *L. welwitschii*.

Shrubs up to 1 m high, glabrous. *Leaves* opposite-subopposite to alternate, mostly elliptical to lanceolate, usually less than 30 mm long, never rounded. *Inflorescence* a few-flowered axillary umbel. *Flowers* 5-merous, gamopetalous, bilaterally symmetrical, cylindrical to subcylindrical in bud, yellow to orange. *Corolla* with deep unilateral split, closed portion of tube usually only a few mm long and slightly swollen; lobes longer than rest of corolla, erect or variously twisted, bent, or reflexed. *Filaments* coiled following the explosive anthesis. *Style* filiform; stigma capitate.

A genus of about 20 species, widespread mostly in the arid regions of tropical Africa.

Odontella welwitschii (*Engl.*) *Balle* in Mitt. bot. StSamml., Münch. 7: 161 (1968); in F.S.W.A. 22: 5 (1969). Type: Angola, along Bero River, *Welwitsch* 4883 (PRE!).

Loranthus welwitschii Engl. in Bot. Jb. 20: 87 (1894); Hiern, Cat. Afr. Pl. Welw. 1, 4: 927 (1900); Sprague in F.T.A. 6, 3: 328 (1910). *Oncocalyx welwitschii* (Engl.) v. Tieghem ex Durand & B. D. Jackson, Ind. Kew, suppl. 1: 301 (1906). *Tapinanthus welwitschii* (Engl.) Danser in Verh. K. Akad. Wet., sect. 2, 29, 6: 122 (1933).

L. elegantissimus Schinz in Bull. Herb. Boissier sér. 1, 4, App. 3: 52 (1896). *Tapinanthus elegantissimus* (Schinz) Danser in Verh. K. Akad. Wet., sect. 2, 29, 6: 111 (1933). Syntypes: South West Africa, Potemine, *Fleck* 415 (Z!); Oombale, *Schinz* 291 (Z!).

L. karibibensis Engl. in Bot. Jb. 40: 524 (1908). Type: South West Africa, Karibib, *Dinter* 1445 (SAM!).

L. engleranus Krause & Dinter in Bot. Jb. 51: 456 (1914). Type: South West Africa, Tsumeb, *Dinter* 1667 (SAM!).

Shrubs to perhaps 1 m high, stems beige to brown. *Leaves* mostly elliptic to lanceolate-oblanceolate, 25–35 × 5–10 mm, usually 3-nerved from just above base, chartaceous to lightly coriaceous; petioles 2–3 mm long. *Inflorescence:* umbels 2–5-flowered; peduncles subsessile to 2 mm long, about as long as pedicels. *Flowers* 20–22 mm long. *Corolla* yellow to yellow-green, expanding laterally at point of filament attachment, lobes c. 12 mm long, usually inflexed to some extent. *Calyx* tubular, 1 mm high. *Berries* unknown. *Flowering* January through March. Fig. 11.

Parasitic on *Acacia* and *Boscia* from central South West Africa north to Angola (Map 8).

Vouchers: *Dinter 6804; De Winter & Leistner 5611; Giess, Volk & Bleissner 5090*.

The combination *Oncocalyx welwitschii* was intended but not validly published by v. Tieghem (1895).

2074e   5. **VANWYKIA**

Vanwykia *Wiens* in Bothalia 12: 422 (1978). Type species: *V. remota* (Bak. & Sprague) Wiens.

Loranthus sect. *Remoti* Sprague in F.T.A. 6, 1: 265 (1910). *L.* sect. *Dendrophthoe* (Mart.) Engl. in Pflanzenfam. ed. 1, 3, 1: 186 (1894), pro parte; Engl. & Krause in Pflanzenfam. ed. 2, 16b: 152 (1935), pro parte quoad 'Gruppe' *Remoti*.

Moderate to large shrubs reaching perhaps 1 m or more high, spreading by haustoria-bearing surface runners. *Stems* stout and robust, leaf-bearing shoots 3–5 mm thick; young branches densely shortly tomentose, older branches glabrate. *Leaves* subopposite-alternate, rarely fascicled, mostly obovate, rounded apically, cuneate into the base, 50–60 × 30–40 mm, densely yellow-white tomentose when young, glabrate and thickly coriaceous with age;

petioles 4–6 mm long. *Inflorescence* a 3–6-flowered umbel, often fascicled on swollen nodes of older branches, completely covered by dense yellowish white tomentum at least 1 mm thick (trichomes with whorls arising from a central axis); peduncles 4–5 mm long. *Flowers* 5-merous, 40 mm long, with short pedicels 1–2 mm long and conspicuous oblong-linear bracts 4–5 mm long. *Corolla* gamopetalous, bilaterally symmetrical (the tube bearing only a short split 4–6 mm deep); lobes erect, somewhat incurved apically. *Filaments* attached at base of corolla lobes, essentially erect, but curving inward to form a central, collective anther mass; anthers 4 mm long. *Style* thickened and pubescent below, thinner and glabrous above; stigma ellipsoid. *Berries* unknown.

A genus with 1 (possibly 2) species in eastern to south-eastern Africa. Related to *Septulina, Bakerella* in Madagascar, and especially to the larger Asian genus *Taxillus*.

Vanwykia remota (*Bak. & Sprague*) *Wiens* in Bothalia 12: 422 (1978). Type: Mozambique, Shupanga, *Kirk* 40 (K!).

L. remotus Bak. & Sprague in F.T.A. 6, 1: 327 (1910).

The single species with characteristics of the genus; *n*=9. Fig. 11.

Parasitic on species of *Lonchocarpus* and *Xeroderris,* especially the latter. Entering Southern Africa near Pafuri in the extreme north-east of Kruger National Park. Known otherwise from Mozambique, Malawi, Zambia and Tanzania (Map 8).

Vouchers: *Van Wyk* 4757; *Wiens & Van Wyk* 5333.

2074f   **6. ERIANTHEMUM**

Erianthemum *v. Tieghem* in Bull. Soc. bot. Fr. 42: 247 (1895); Danser in Verh. K. Akad. Wet., sect. 2, 29, 6: 55 (1933); Balle in F. S. W. A. 22: 4 (1968). Type species: *E. dregei* (Eckl. & Zeyh.) v. Tieghem.

Loranthus subgen. *Erianthemum* (v. Tieghem) Balle in F.C.B. 1: 308 (1948).

L. sect. *Dendrophthoe* (Mart.) Engl. in Pflanzenfam. ed. 1, 3, 1: 186 (1894), pro parte; in Bot. Jb. 20: 104 (1894), pro parte quoad partem 'Gruppe' *Hirsutorum. L.* sect. *Hirsuti* Engl. ex Sprague in F.T.A. 6,1: 263 (1910); in Kew Bull. 1915: 70 (1915).

Moderate to large shrubs up to 2 m high, younger stems and leaves densely brownish to whitish stellate-pubescent. *Leaves* opposite-subopposite, sometimes crowded in fascicles on older stems. *Inflorescence* a 2–6-flowered head, axillary or terminating short spur-branches originating from leaf fascicles on older stems. *Flowers* 5-merous with conspicuous, whitish, long (3–5 mm), ascending, silky hairs. *Corolla* radially symmetrical at anthesis, without unilateral split, lobes variously reflexed at anthesis, usually longer than or equalling tube. *Filaments* attached well above base of lobes; anthers breaking from filaments with explosive opening of corolla, remaining filaments essentially erect. *Style* filiform; stigma capitate. *Berries* reddish, with long silky hairs, calyx tube persistent.

A genus of about 15 species widely distributed from central to southern Africa.

Leaves at maturity usually glabrous, at least on upper surface, usually borne singly on new branches, generally exceeding 40 × 20 mm; young stems and petioles usually rusty pubescent . . . . 1. *E. dregei*

Leaves at maturity evenly stellate-pubescent on both surfaces, borne in dense fascicles mostly on older branches, generally less than 30 × 15 mm; young stems and petioles usually whitish pubescent . . . . 2. *E. ngamicum*

1. Erianthemum dregei (*Eckl. & Zeyh.*) *v. Tieghem* in Bull. Soc. bot. Fr. 42: 247 (1895). Type: Cape, Bothasberg, *Ecklon & Zeyher* 2284 (K!; SAM!).

L. dregei Eckl. & Zeyh., Enum. 358 (1837); Harv. in F.C. 2: 575 (1862); Wood & Evans, Natal Plants 4, pl. 312 (1903); Sprague in F.T.A. 6, 1: 311 (1910); in F.C. 5, 2: 109 (1915); Burtt Davy, Fl. Transv. 465 (1932); Batten & Bokelmann, Wild Flow. E. Cape Prov., pl. 55, 3 (1966); Ross, Fl. Natal 152 (1972); Gibson, Wild Flow. Natal, pl. 30, 2 (1975).

L. dregei forma *subcuneifolia* Engl. in Bot. Jb. 20: 104 (1894). Type: Cape, *Drège* 2284 (SAM, lecto.!).

L. dregei forma *obtusifolia* Engl. in Bot. Jb. 20: 105 (1894). Syntypes.

L. oblongifolius E. Mey. ex Drège in Zwei Pfl. Doc. 148, 159 (1844), nom. nud.

Large shrubs, probably up to 2 m high, young parts usually brownish pubescent. *Leaves* mostly glabrous at maturity, usually oblong, size highly variable, 40–100 × 20–30 mm, lateral venation conspicuous; petioles 6–10 mm long. *Inflorescence:* heads, 1–3, axillary; peduncles 8–15 mm long, often brownish pubescent; bracts 3 × 2 mm, half or less the length of the ovary and calyx tube. *Berries* red, 11–12 mm long. *Flowering* June–March, and possibly throughout the year in differing climatic regions; $n=9$. Fig. 12.

Parasitic on a great diversity of hosts including species of *Acacia, Brachylaena, Chrysophyllum, Combretum, Euclea, Grewia, Hymenospermum, Kiggelaria, Maytenus, Melia, Rhus, Schotia, Spirostachys, Strychnos, Trichilia,* and no doubt many other genera. Widely distributed throughout the Transvaal, the coastal regions and midlands of Natal, Transkei and the eastern Cape (Map 9). Also apparently widespread in tropical Africa.

Vouchers: *Codd* 4400; *Galpin* 9505; *Strey* 5266.

Perhaps the most ubiquitous and variable of the loranthaceous mistletoes in Southern Africa. The species needs considerable study to determine if some of the highly varied forms require taxonomic separation. One of these is a densely brownish pubescent form from the Soutpansberg and also the Barberton area. This form appears to approach some of the characteristics of *E. schelei* (Engl.) v. Tieghem, a possibly distinct species of tropical Africa. Until more is known of the consistency and geographical distribution of the characters, it seems best to recognize only a single, polymorphic species.

2. Erianthemum ngamicum (*Sprague*) *Danser* in Verh. K. Akad. Wet., sect. 2, 29, 6: 54 (1933). Syntypes: Botswana, Ntschokutsa, *Seiner* 2/124 (Z); Kwebe Hills, *Mrs Lugard* 44 (K!); Lake Ngami, *E. J. Lugard* 30 (K!); *Fleck* 313A.

Loranthus ngamicus Sprague in F.T.A. 6, 1: 310 (1910).

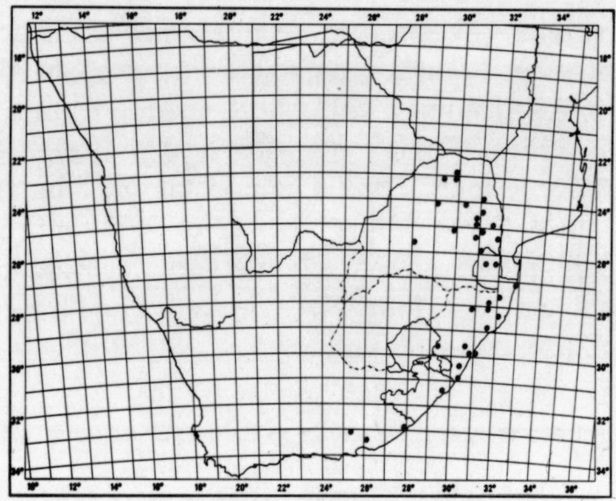

MAP 9.—**Erianthemum dregei**

Shrubs up to 1 m high, young parts usually with dense whitish stellate hairs. *Leaves* opposite-subopposite on young stems, but crowded in fascicles on the often swollen nodes of older stems, mostly elliptic, 15–25 × 10–15 mm, lateral venation inconspicuous; petioles 4–8 mm long. *Inflorescence:* heads 3–5-flowered, arising singly and centrally from crowded leaf fascicles; peduncles often whitish pubescent, 15–25 mm long; calyx rim often with minute but distinct tufts of hairs; bracts at least half or more the length of ovary and calyx tube. *Berries* red, 13–15 mm long. *Flowering* October to March and possibly longer. Fig. 12.

Parasitic on species of *Acacia, Albizia, Colophospermum, Combretum, Commiphora, Pteroxylon, Sclerocarya,* and probably also other genera; from north-eastern South West Africa, Botswana, northern and central Transvaal to northern Natal (Map 10). Also in Rhodesia and Mozambique.

Vouchers: *De Winter* 3723; *Acocks* 13900; *Galpin* 13990.

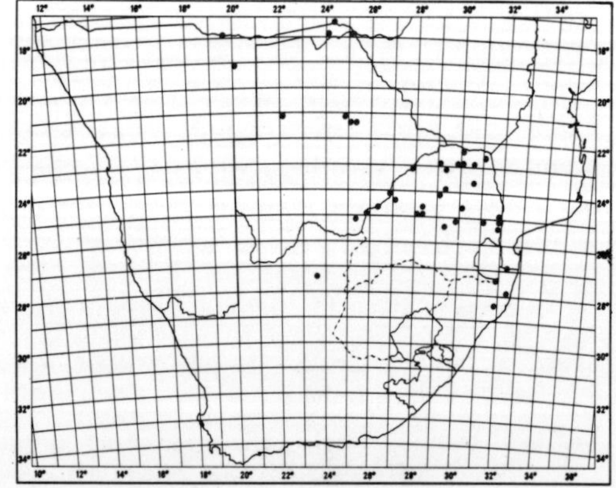

MAP 10.—**Erianthemum ngamicum**

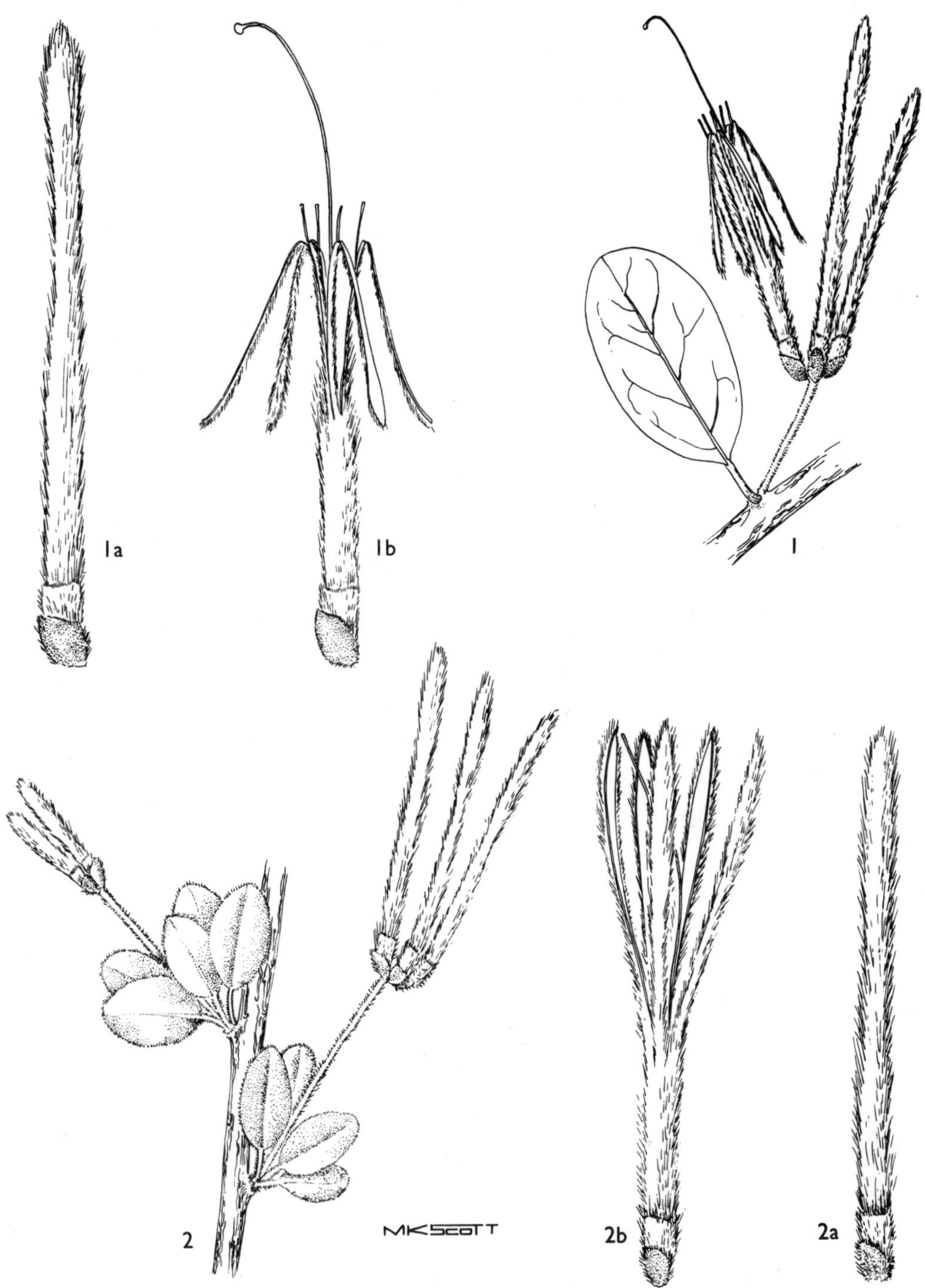

FIG. 12.—1, **Erianthemum dregei,** flowering twig, ×1; 1a, mature bud, ×1,5; 1b, flower, ×1,5 (*Ward* 6438).
2, **E. ngamicum,** flowering twig, ×1; 2a, mature bud, ×1,5; 2b, flower, ×1,5 (*Allen* 203).

2074g   7. **PEDISTYLIS**

Pedistylis *Wiens* in Bothalia 12: 421 (1978). Type species: *P. galpinii* (Schinz ex Sprague) Wiens.

Loranthus sect. *Dendrophthoe* (Mart.) Engl. in Pflanzenfam. ed. 1, 3, 1: 186 (1894), pro parte; Krause in Pflanzenfam. ed. 1, Nachtr. 4: 72 (1914), pro parte quoad 'Gruppe' *Tetrameri*; Engl. & Krause in Pflanzenfam. ed. 2, 16b: 152 (1933), pro parte quoad 'Gruppe' *Tetrameri*. *L.* sect. *Tetrameri* Engl. ex Sprague in Kew Bull. 1915: 69 (1915).

Emelianthe Danser in Verh. K. Akad. Wet., sect. 2, 29, 6: 54 (1933), pro parte quoad *E. galpinii*.

Large shrubs perhaps exceeding 2 m high, glabrous, older plants forming much enlarged (up to 1 m across!) haustorial connections with host. *Younger branches* often densely lenticellate, nodes of older branches usually greatly swollen. *Leaves* opposite-subopposite, mostly oblong, often slightly falcate, highly variable in length (50–) 70–80 (–120) × 10–20 mm, penninerved, veins often raised on lower surface; petioles 8–10 mm long. *Inflorescence* a 2(–3)-flowered umbel, axillary, or often in pairs on swollen nodes of older branches, with stout peduncles and pedicels, each about 5 mm long. *Flowers* 5-merous, radially symmetrical (tube without unilateral split). *Corolla* yellow, gamopetalous, lobes reflexed and about as long as tube; mature buds 70–80 mm long. *Filaments* at maturity curving outward about 90 degrees or more, becoming reddish, attached at base of lobes. *Style* near apex bending downward in a broad curve for 180 degrees or more (as a shepherd's crook), also becoming reddish; stigma broadly ovate. *Berries* ellipsoid, c. 20 × 12 mm, yellow-green, with scattered warts. *Flowering* from approximately February through April; $n=9$.

A monotypic genus narrowly restricted to the lowveld of south-eastern Transvaal and adjoining Swaziland and Mozambique. An isolated genus without apparent close relatives.

Pedistylis galpinii (*Schinz ex Sprague*) *Wiens* in Bothalia 12: 422 (1978). Type: Transvaal, near Barberton, *Galpin* 896 (K, holo.!; PRE!; SAM!).

Loranthus galpinii Schinz ex Sprague in F.C. 5, 2: 112 (1915); Burtt Davy, Fl. Transv. 465 (1932). *Emelianthe galpinii* (Schinz ex Sprague) Danser in Verh. K. Akad. Wet., sect. 2, 29, 6: 53 (1933).

The single species with characteristics of the genus. Fig. 13.

Parasitic on species of *Acacia, Combretum, Dichrostachys, Sclerocarya, Terminalia* and *Trichilia*. Known from south-western Kruger National Park and the lower elevations of Swaziland. Also in adjoining Mozambique (Map 11).

Vouchers: *Codd* 6049; *Evans* 3469; *Galpin* 896.

This species was first recognized as a distinct genus by Danser. Unfortunately, Danser followed Sprague who believed it was closely related to an east African species then known as *Loranthus panganensis* Engl. Danser included the two species in his genus *Emelianthe*, naming *E. panganensis* (Engl.) Danser as the type. Pressed specimens of these species show similarities in size and form but the resemblance is superficial. In fact, field studies show that the species have little in common beyond the actinomorphic corolla and the overall size of the flower and they are not closely related.

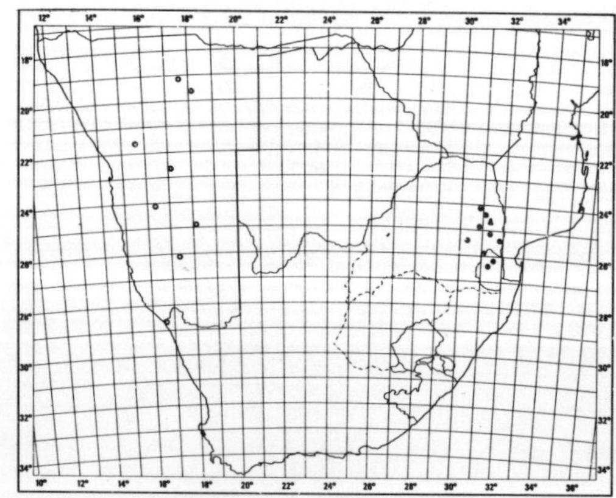

MAP 11.—●Pedistylis galpinii
○Plicosepalus undulatus
▲Plicosepalus amplexicaulis

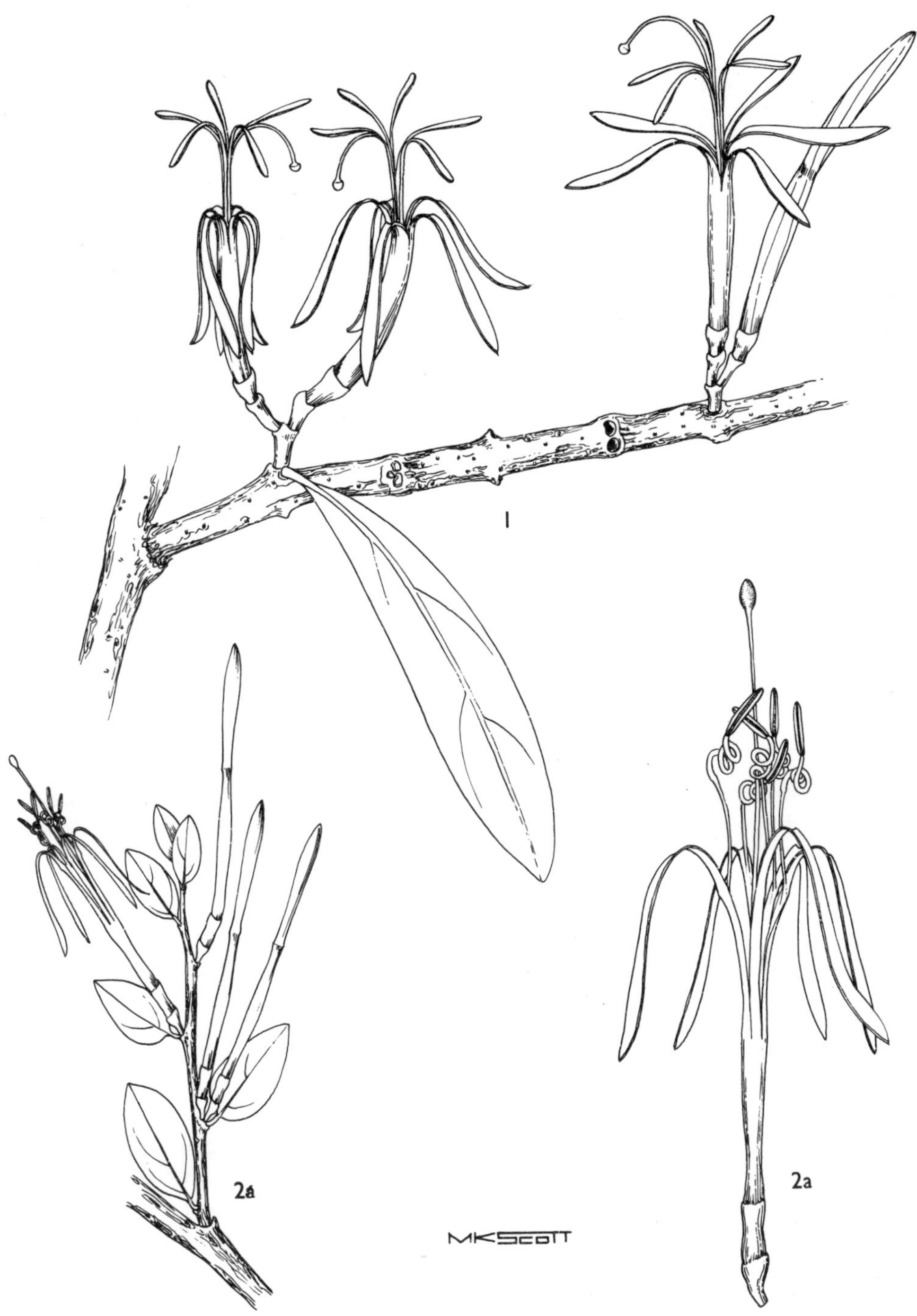

FIG. 13.—1, **Pedistylis galpinii**, flowering twig, ×0,7 (*Wiens* 5257). 2, **Actinanthella wyliei**, flowering twig, ×1; 2a, flower, ×2 (*Huntley* 894).

2074h   8. ACTINANTHELLA

Actinanthella *Balle* in Bull. Séanc. Inst. r. colon. belge 25: 1625 (1954). Type species: *A. menyharthii* (Engl. & Schinz) Balle.

Loranthus sect. *Dendrophthoe* (Mart.) Engl. in Pflanzenfam. ed. 1, 3, 1: 186 (1894), pro parte; Krause in Pflanzenfam. ed. 1, Nachtr. 4: 72 (1914), pro parte quoad "Gruppe" *Incrassati*. *L.* sect. *Incrassati* Krause ex Sprague in F.T.A. 6, 1: 263 (1910); in Kew Bull. 1915: 70 (1915).

Shrubs up to perhaps 1 m high, stems puberulent to densely pubescent. *Leaves* alternate, rarely in small fascicles, oblong-oblanceolate to elliptical, glabrous or pubescent; petioles 2–3 mm long. *Flowers* 5-merous, solitary, sessile-subsessile, 1–2 in axils. *Corolla* yellow-green, sometimes with red, gamopetalous, radially symmetrical (without unilateral split) tube expanding upward to at least twice the basal diameter, lobes 2–3 times as long as tube. *Filaments* attached slightly below midpoint of lobe, distal portion only coiled following anthesis, coiled portion approximately twice as wide and thick as proximal, erect portion. *Style* filiform; stigma ovoid; $n=9$.

A genus of 2 species endemic to south-eastern and Southern Africa; without obvious close relatives.

Actinanthella wyliei (*Sprague*) *Wiens* in Bothalia 12: 423 (1978). Type: Natal, Ngoya Forest Reserve, *Wylie* sub Wood 7468 (K, holo.!; PRE!).

Loranthus wyliei Sprague in Kew Bull. 1915: 78 (1915); in F.C. 5, 2: 110 (1915); Ross, Fl. Natal 153 (1972). *Tapinanthus wyliei* (Sprague) Danser in Verh. K. Akad. Wet., sect. 2, 29, 6: 122 (1933).

Small, densely branched shrubs, probably less than 0,5 mm high, glabrous except for the puberulent young stems. *Internodes* mostly short, but also variable (3–12 mm long). *Leaves* irregularly alternate, sometimes in small fascicles, oblong-oblanceolate, 10–15×5–10 mm; petioles 2 mm long. *Flowers* yellow-green basally, pinkish red above, 35–40×1–2 mm; lobes variously reflexing at anthesis. *Calyx* tubular, 2–3 times longer than ovary. *Berries* unknown. *Flowering* February and March (and probably also earlier in the season). Fig. 13.

Parasitic on *Erythroxylum* sp. and known only from the Ngoya Forest Reserve in central Natal.

Vouchers: *Garland* 346; *Huntley* 894.

An apparently rare, or at least little collected species. Related to *A. menyharthii* of Rhodesia and Mozambique, the only other species in the genus.

2074i   9. PLICOSEPALUS

Plicosepalus *v. Tieghem* in Bull. Soc. bot. Fr. 41: 504 (1894); Danser in Verh. K. Akad. Wet., sect. 2, 29, 6: 100 (1933). Type species: *P. undulatus* (Harv.) v. Tieghem.

Loranthus sect. *Plicopetalus* Benth. & Hook. f., Gen. Pl. 3, 1: 208 (1880); Engl. in Pflanzenfam. ed. 1, 3, 1: 188 (1894); in Bot. Jb. 20: 130 (1894); Sprague in F.T.A. 6, 1: 258 (1910); in Kew Bull. 1915: 69 (1915). *L.* sect. *Plicotepalus* Engl. & Krause in Pflanzenfam. ed. 2, 16b: 148 (1935), nom. superfl.

Shrubs reaching 1 m or more high, glabrous, spreading by haustoria-bearing surface runners. *Leaves* opposite-subopposite, mostly oblong to elliptical, highly coriaceous with age, shortly petiolate. *Inflorescence* an axillary, solitary or fascicled, 2–6-flowered umbel. *Flowers* 5-merous, bilaterally symmetrical (strongly curved, especially in older buds, forming a "C"- or bow-shaped), base rather enlarged and bud apex strongly clavate. *Corolla* choripetalous, but connivent and superficially appearing united at base, usually red or yellow. *Petals* basally plicate on inner surface, lobes variously reflexing (often twisted) above point of filament attachment, remaining connivent below. *Calyx* reduced to a short rim, rarely over 1 mm high. *Style* filiform, inserted in a depression on ovary; stigma capitate.

A genus of perhaps 10 species, widespread throughout the more arid regions of Africa.

Leaves basally rounded to cuneate, subpetiolate-petiolate, mostly oblong-linear to elliptic, occasionally falcate, if in dense fascicles along stem, then mostly obovate:

Flowers pinkish red; style curved or with single basal bend; berries red, smooth .....1. *P. kalachariensis*

Flowers yellow-orange; style with double basal bend ("S"-shaped); berries yellow, warty ...2. *P. undulatus*

Leaves conspicuously sagittate, clasping stem, linear to linear-oblong ....3. *P. amplexicaulis*

1. **Plicosepalus kalachariensis** (*Schinz*) *Danser* in Verh. K. Akad. Wet., sect. 2, 29, 6: 100 (1933). Syntypes: Botswana, Okavango, *Fleck* 307 (not traced); Lake Ngami, *Fleck* 314a (not traced; in Z fide N.E. Br. in lit., 25/11/1909).

Loranthus kalachariensis Schinz in Bull. Herb. Boissier, sér. 1, 4, App. 3: 53 (1896); Sprague in F.T.A. 6, 1: 280 (1910); in F.C. 5, 2: 105 (1915); Burtt Davy, Fl. Transv. 465 (1932).

L. dinteri Schinz in Bull. Herb. Boissier, sér. 2, 1: 869 (1901). Type: South West Africa, Grootfontein, *Dinter* 698 (not traced; "identical with type of L. kalachariensis" fide N.E. Br. in lit., 25/11/1909).

L. splendens N.E. Br. in Kew Bull. 1909: 136 (1909). Type: Botswana, Okavango, *Lugard* 232 (K!). *L. acaciaedetinentis* Dinter in Reprium nov. Spec. Regni veg. 18: 441 (1922). *Plicosepalus acaciaedetinentis* (Dinter) Danser in Verh. K. Akad. Wet., sect. 2, 29, 6: 100 (1933). Syntypes: South West Africa, Etosha Pan, *Dinter* 2265 (SAM!); Grootfontein and Outjo, *Dinter* s.n. (not traced).

Plicosepalus curviflorus sensu Balle in F.S.W.A. 22: 8 (1969), non (Benth.) v. Tieghem.

Shrubs of moderate to large size, often 1 m or higher. *Leaves* oblong-linear to elliptic (highly variable in size), (25–) 40–60 (–85)× 7–20 mm, basinerved, the 3–5 veins sometimes faint. *Inflorescence:* umbels mostly 2–4 (–6)-flowered, borne singly or in pairs, axillary, often on swollen nodes of older branches; pedicels 8–11 mm long. *Corolla* pink to reddish orange, usually darker toward base, colours often uneven, mottled, or variegated. *Filaments* and style pinkish apically. *Style* with single bend toward base. *Berries* red, smooth. *Flowering* at least April through August and probably longer; $n=9$. Fig. 14.

Parasitic apparently only on *Acacia* species. Widespread in the northern half of South West Africa, in northern Botswana, northern and eastern Transvaal as well as Swaziland and northern Natal (Map 12).

Vouchers: *Codd* 8725; *Dinter* 7690; *Galpin* 14879.

This species is part of a widespread complex ranging southward from north-eastern Africa. Balle (F.S.W.A. 22: 8) placed this species in synonymy under *P. curviflorus* (Benth.) v. Tieghem. Field studies in both east and Southern Africa, however, show a number of differences between these population systems. Furthermore the species distinctions in the genus in tropical Africa are not clear, and until more is known of these inter-relationships the established name for this taxon is retained.

2. **Plicosepalus undulatus** (*E. Mey. ex Harv.*) *v. Tieghem* in Bull. Soc. bot. Fr. 41: 504 (1894); Balle in F.S.W.A. 22: 8 (1969). Syntypes: Cape, between Holgat and Orange River, *Drège* s.n. (K!; PRE!); between Verleptpram and Orange River mouth, *Drège* s.n. (K!).

Loranthus undulatus E. Mey. ex Harv. in F.C. 2: 577 (1862); Schinz in Bull. Herb. Boissier, sér. 1, App. 4: 54 (1896); Sprague in F.T.A. 6, 1: 278 (1910); in F.C. 5, 2: 104 (1915).

L. fleckii Schinz in Bull. Herb. Boissier, sér. 1, 4, App. 3: 53 (1896). Syntypes: South West Africa, Ubib, *Fleck* 416 (Z); Potemine, *Fleck* 404 (Z).

L. undulatus var. *angustior* Sprague in F.C. 5, 2: 105 (1915). Type: South West Africa, Sandverhaar, *Pearson* 4694 (K, holo.!).

This species differs from *P. kalachariensis* by the often dimorphic leaves, which maybe oblong-linear and usually over 20 mm long, or densely fascicled, broadly ovate, less than 20 mm long, and often subtending inflorescences. *Inflorescence:* umbels mostly 2-flowered; bracts linear-oblong, 3–4 mm long, broadly keeled. *Corolla* yellow-orange, petals often minutely ribbed at margin. *Style* with double bend near base. *Berries* yellow, warty. Fig. 14.

Parasitic primarily on species of *Acacia* (also on *Terminalia*) from the north-western Cape Province to northern South West Africa (Map 11).

Vouchers: *De Winter* 2344; *Dinter* 285; 306; *Kinges* 2078; 3240.

3. **Plicosepalus amplexicaulis** *Wiens* in Bothalia 12: 422 (1978). Type: Transvaal, Kruger National Park, Balule Camp, *Wiens* 4681 (K, holo.!; PRE!; UT!).

Moderate-sized shrubs up to about 1 m high. *Branches* buff to brown; internodes 30–40 mm. *Leaves* sessile, linear to linear-oblong, 40–50×6–10 mm, usually 3-nerved, light grey-green, basal lobes sagittate, 6–8 mm long, amplexicaul. *Inflorescence:* umbels 3–4-flowered, axillary, borne singly or in pairs; peduncles and pedicels approximately equal, 7–9 mm long. *Corolla* at anthesis 45–50 mm long, dull white basally; lobes bright red,

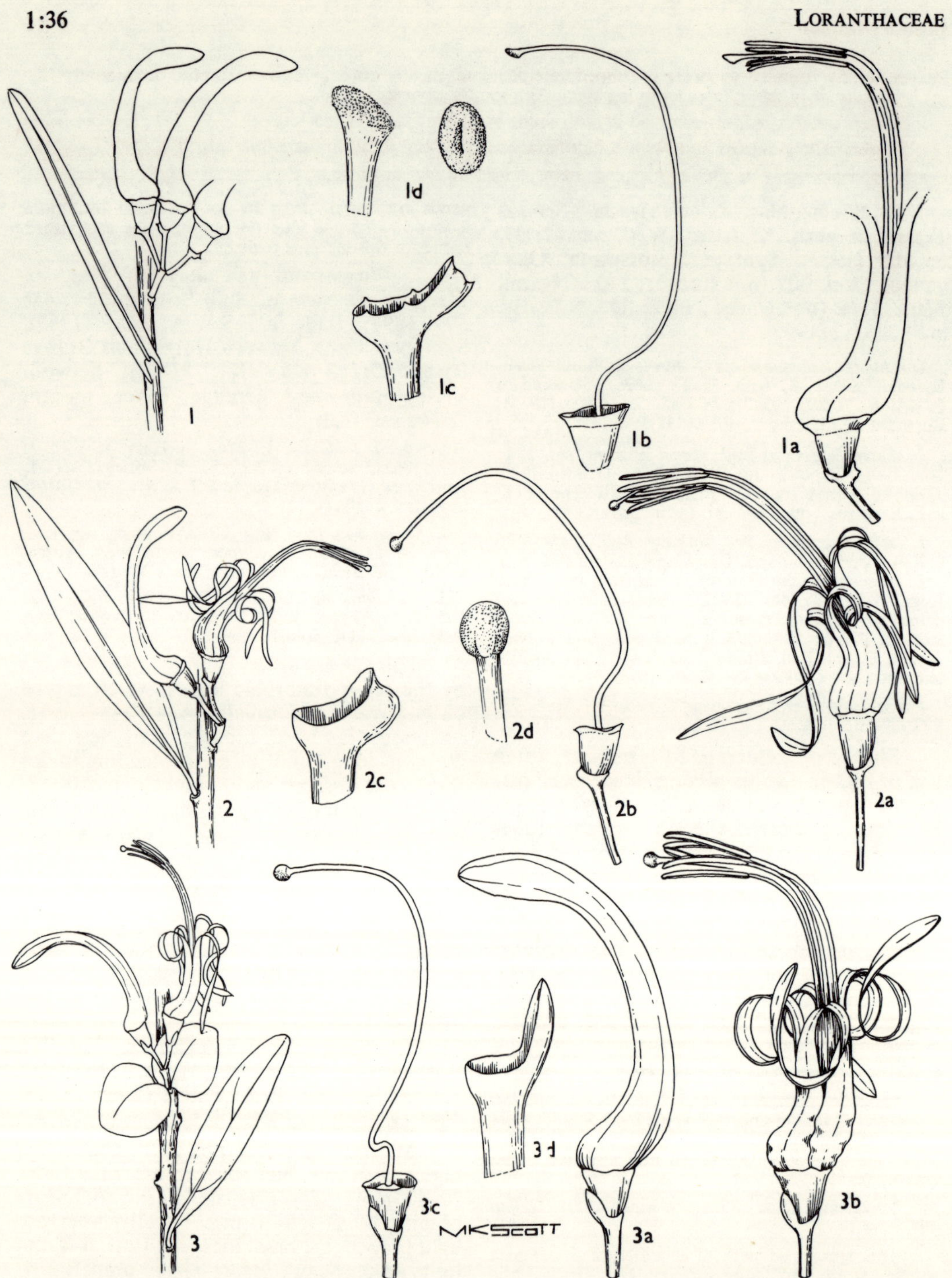

Fig. 14.—1, **Plicosepalus amplexicaulis**, flowering twig, ×0,7; 1a, flower, ×1,5; 1b, style, ×1,5; 1c, bract, ×6; 1d, stigma, ×6 (*Van der Schijff* 260). 2, **P. kalachariensis**, flowering twig, ×0,7; 2a, flower, ×1,5; 2b, style, ×1,5; 2c, bract, ×6; 2d, stigma, ×6 (*Werger* 994). 3, **P. undulatus**, flowering twig, ×0,7; 3a, mature bud, ×1,5; 3b, flower, ×1,5; 3c, style, ×1,5; 3d, bract, ×6 (*Leach & Bayliss* 13005).

(dull purplish in older buds). *Filaments, style* and *stigma* bright red (as the lobes); style with a single basal bend. *Berries* ellipsoid at maturity, 11×7 mm. *Flowering* in June and July (possibly longer); $n=9$. Fig. 14.

A parasite on *Acacia* species, apparently restricted to the lowveld of the eastern Transvaal in the Kruger National Park and adjoining areas (Map 11).

Vouchers: *Wiens* 4681; *Van der Schijff* 260; *Marassas* 813.

This species is clearly distinguished from *P. kalachariensis* by the amplexicaul, sagittate leaves and the basally whitish corolla. The species is also distinct from *P. sagittifolius* of eastern and north-eastern Africa which has a differently coloured corolla and differently shaped, often fascicled leaves.

2074j

10. HELIXANTHERA

Helixanthera *Lour.*, Fl. Cochinch. 1: 142 (1790); Danser in Verh. K. Akad. Wet., sect. 2, 29, 6: 55 (1933). Type species: *H. parasitica* Lour.

Loranthus sect. *Acrostachys* Benth. & Hook. f., Gen. Pl. 3, 1: 208 (1880); Engl. in Pflanzenfam. ed. 1, 3, 1: 188 (1894), pro parte; Nachtr. 1: 133 (1897), pro parte; Sprague in F.T.A. 6, 1: 258 (1910), pro parte; in Kew Bull. 1915: 69 (1915), pro parte; Engl. & Krause in Pflanzenfam. ed. 2, 16b: 147 (1935), pro parte quoad *L. garcianus*. *Achrostachys* (Benth. & Hook. f.) v. Tieghem in Bull. Soc. bot. Fr. 41: 504 (1894).

Sycophila Welw. ex v. Tieghem in Bull. Soc. bot. Fr. 41: 485 (1894). *L.* sect. *Sycophila* (Welw. ex v. Tieghem) Engl., Pflanzenfam., Nachtr. 1: 128 (1897); Sprague in Kew Bull. 1915: 69 (1915); Engl. & Krause in Pflanzenfam. ed. 2, 16b: 147 (1935), pro parte quoad *L. woodii*.

Small to moderate-sized woody or succulent shrubs up to perhaps 0,5 m high, glabrous with mostly greyish to brownish stems. *Leaves* opposite-subopposite, broad to elongate, succulent to chartaceous. *Inflorescence* a terminal or axillary raceme, 15–40 mm long, 8–40-flowered. *Flowers* 4-merous, $5–15 \times 1–3$ mm, whitish to reddish-orange; pedicels short (1–3 mm). *Corolla* choripetalous, lobes reflexing from point of filament attachment. *Filaments* erect; anthers with or without locellae. *Style* quadrangular or sometimes thickened basally. *Berries* red to pink.

A genus of about 50 species, with perhaps 10 species in Africa south of the Sahara and the remaining in tropical Asia. Several Asian species are shrubby, terrestrial root parasites.

Inflorescence stout and elongated, rachis 100–150 mm long, 2–3 mm wide; petioles 10–12 mm long; leaves thick and succulent .... 1. *H. garciana*

Inflorescence delicate and short, rachis 30–40 mm long, c. 1 mm wide or less; petiole 2–5 mm long; leaves thin and chartaceous:

Leaves 30–50 mm long, margins smooth and even; buds at maturity c. 12 mm long with basal ribs extending to the calyx rim; style after anthesis c. 9 mm long .... 2. *H. subcylindrica*

Leaves 20–30 mm long, margins undulate; buds at maturity c. 7 mm long with basal ribs limited to a small area about 3 mm above base; style after anthesis c. 5 mm long................. 3. *H. woodii*

1. Helixanthera garciana (*Engl.*) *Danser* in Verh. K. Akad. Wet., sect. 2, 29, 6: 57 (1933). Type: Mozambique, Ressano-Garcia, *Schlechter* 11921 (K!).

Loranthus garcianus Engl. in Bot. Jb. 40: 539 (1908); Sprague in F.C. 5, 2: 103 (1915); Burtt Davy, Fl. Transv. 465 (1932).

L. messinensis N.E. Br. in Kew Bull. 1921: 296 (1921). *Helixanthera messinensis* (N.E. Br.) Danser in Verh. Akad. Wet., sect. 2, 29, 6: 58 (1933). Type: Transvaal, Messina, *Rogers* 22568 (K, holo.!; PRE!).

Robust, highly succulent shrubs. *Terminal branchlets* with stout internodes, $20–30 \times 2–3$ mm, new growth often reddish brown. *Leaves* rounded to ovate-obovate, 30–50 $\times$ 15–30 mm, thickly succulent (coriaceous when dried). *Racemes* terminal, 100–150 mm long, surpassing the leaves, 20–40-flowered. *Flowers* reddish orange; pedicels 3 mm long. *Berries* rounded to ellipsoid, bright red, c. 10 mm high. *Flowering* from approximately December through February. Fig. 15.

Parasitic on *Sclerocarya caffra* Sond. in the northern and north-eastern Transvaal (Map 12).

Vouchers: *Rogers* 18978; *Werdermann & Oberdieck* 2004; *Wiens* 5318.

Closely related and possibly conspecific with *H. kirkii* (Oliv.) Danser, a species widely distributed in central and eastern Africa.

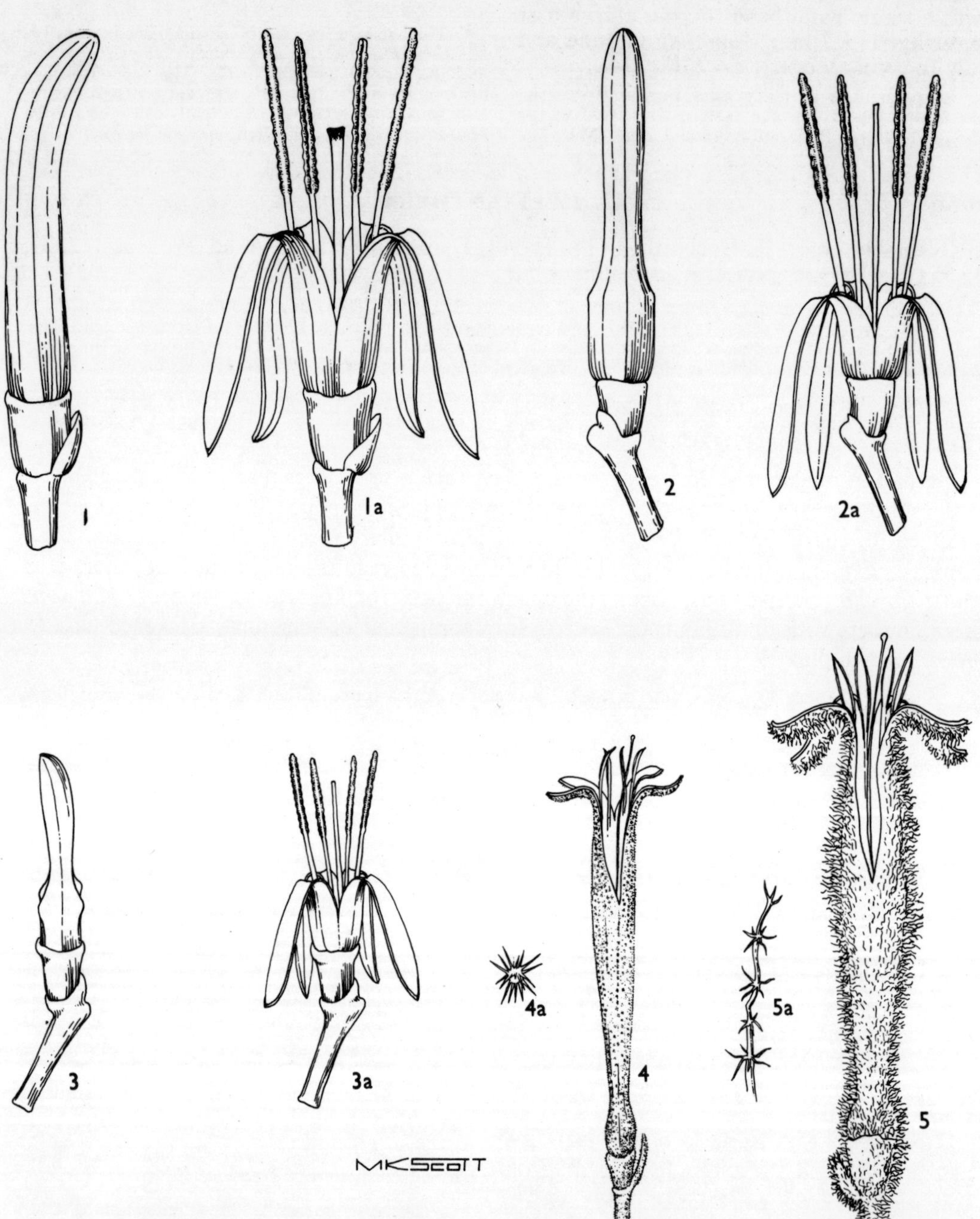

FIG. 15.—1, **Helixanthera garciana**, mature bud, ×4; 1a, flower, ×4 (*Werdermann & Oberdieck* 2004). 2, **H. subcylindrica**, mature bud, ×4; 2a, flower, ×4 (*Strey* 9592). 3, **H. woodii**, mature bud, ×4; 3a, flower, ×4 (*Huntley* 905). 4, **Septulina glauca**, flower, ×2; 4a, hair, ×30 (*Compton* 24179). 5, **S. ovalis**, flower, ×2; 5a, hair, ×30 (*Van Son* in TRV 36614).

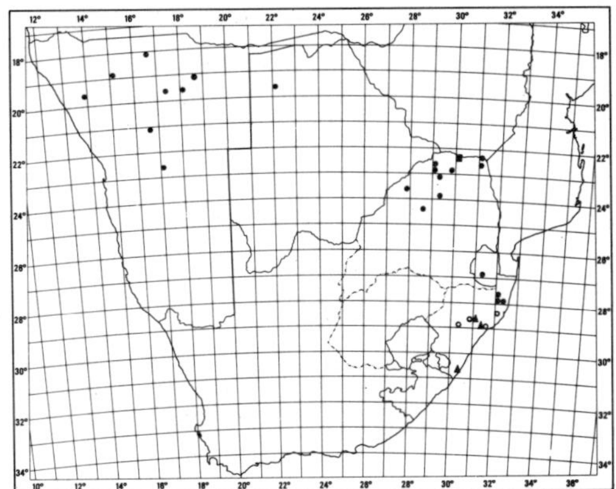

MAP 12.—●Plicosepalus kalachariensis
■Helixanthera garciana
▲H. subcylindrica
○H. woodii

2. Helixanthera subcylindrica (*Sprague*) *Danser* in Verh. K. Akad. Wet., sect. 2, 29, 6: 59 (1933). Syntypes: Natal, Alexandra, *Rudatis* 904 (K!; PRE!; Z!); Nkandla, *Wylie* sub Wood 9013 (K!).

Loranthus subcylindricus Sprague in Kew Bull. 1915: 78 (1915); in F.C. 5, 2: 103 (1915); Ross, Fl. Natal 152 (1972).

L. woodii Schltr. & Krause in Bot. Jb. 51: 454 (1914), pro parte quoad *Rudatis* 904.

Open, rather laxly branched shrubs. *Branchlets* with internodes mostly 20–35×1 mm. *Leaves* lanceolate-ovate to elliptical, 30–40×10–15 mm, margins essentially smooth and flattened, chartaceous. *Inflorescence:* raceme terminal, 20–30 mm long, 10–20-flowered. *Flowers* whitish, 10×1 mm at anthesis; pedicels 2 mm long. *Berries* ellipsoid, 8–9 mm long, pink. *Flowering* approximately from December through February. Fig. 15.

Parasitic on *Ochna* spp. and probably other hosts in the central and southern midlands of Natal (Map 12).

Vouchers: *Edwards* 1432; *De Winter* 8257; *McClean* 360.

3. Helixanthera woodii (*Schltr. & Krause*) *Danser* in Verh. K. Akad. Wet., sect. 2, 29, 6: 60 (1933). Type: Natal, Ngoya, *Wood* 3874 (K!; PRE!).

Loranthus woodii Schltr. & Krause in Bot. Jb. 51: 454 (1914), pro parte excl. *Rudatis* 904; Sprague in F.C. 5, 2: 102 (1915); Ross, Fl. Natal 152 (1972); Gibson, Wild Flow. Natal, pl. 30, 6 (1975).

L. sandersonii Harv. ex Benth. & Hook.f., Gen. Pl. 3: 208 (1880), nom. nud. *Acrostachys sandersonii* v. Tieghem in Bull. Soc. bot. Fr. 41: 504 (1894), nom. nud.

Compact, much branched shrubs. *Branchlets* with internodes mostly 10–15×1 mm. *Leaves* ovate-lanceolate to oblong-elliptical, 15–25×10–15 mm, margins strongly undulate, chartaceous. *Inflorescence:* raceme axillary or terminal, 15–20 mm long, 8–15-flowered. *Flowers* whitish, 5×0,5 mm at anthesis; pedicels c. 1 mm long. *Berries* ellipsoid, 5–6 mm long, scarlet. *Flowering* approximately from December through February. Fig. 15.

Parasitic on *Burchellia bubalina* (L.f.) Sims and probably also other hosts; known only from the Ngoya Forest Reserve in central Natal (Map 12).

Vouchers: *Huntley* 905; *Wells & Edwards* 107; *Wood* 9402.

2074k   **11. SEPTULINA**

Septulina *v. Tieghem* in Bull. Soc. bot. Fr. 42: 263 (1895); Balle in Mitt. bot. StSamml., Münch. 7: 177 (1968); in F.S.W.A. 22: 9 (1969). Type species: *S. glauca* (Thunb.) v. Tieghem.

Loranthus sect. *Septulina* (v. Tieghem) Sprague in Kew Bull. 1914: 367 (1914); in Kew Bull. 1915: 69 (1915).

Scurrula L., Sp. Pl. ed. 1, 1: 110 (1753), nom. rejic.; G. Don, Gen. Hist. Dichlam. 3: 401, 423 (1834), pro parte quoad *S. canescens* et *S. glaucus*. L. sect. *Scurrula* (L.) DC., Prodr. 4: 301 (1830); Coll. Mem., Loranth. 6 (1830), pro parte.

Dendrophthoe Mart. in Flora 13: 109 (1830), pro parte quoad *D. glaucus*. L. sect. *Dendrophthoe* (Mart.) Engl., Pflanzenfam. ed. 1, 3, 1: 186 (1894); in Bot. Jb. 20; 81 (1894), pro parte quoad partes 'Gruppe' *Cinerascentorum*; Engl. & Krause in Pflanzenfam. ed. 2, 16b: 152 (1935), pro parte quoad partes 'Gruppe' *Cinerascentorum*. L. subgen. *Dendrophthoe* (Mart.) Engl., Pflanzenfam., Nachtr. 1: 129 (1897), pro parte quoad partes 'Gruppe' *Cinerascentorum*.

L. sect. *Cichlanthus* Endl. ex Benth. & Hook. f., Gen. Pl. 3, 1: 209 (1880), pro parte quoad *L. glaucus* et *L. ovalis*.

Taxillus v. Tieghem in Bull. Soc. bot. Fr. 42: 256 (1895); Danser in Verh. K. Akad. Wet., sect. 2, 29, 6: 123 (1933), pro parte quoad *T. glaucus* et *T. ovalis*; Balle in Webbia 11: 580 (1955).

Shrubs of small to moderate size, probably not higher than 0,75 m, often appearing greyish by the presence of short, dense tomentum on stems and leaves. *Older stems* often greyish to reddish brown. *Leaves* alternate, shortly petiolate (2–6 mm), obovate-elliptical, 10–30 × 3–20 mm. *Flowers* 4-merous, solitary, 1–3 in axils with short pedicels (1–3 mm). *Corolla* stellate or lanate-pubescent, 25–40 mm long, with short unilateral split 5–7 mm long, outside mostly greyish green with variable shades of dull red, inside orange-reddish, lobes reflexed. *Calyx* reduced to a short rim or forming a tube up to 2 mm long; bracts 2–3 mm long. *Filaments* more or less erect, 4–6 mm long; anthers locellate, about as long as filaments. *Style* essentially filiform but tapering slightly towards stigma.

A genus of only 2 species confined to South Africa and western Cape Province. Related to *Vanwykia* and the Madagascan genus *Bakerella*, both of which show close affinities to the larger genus *Taxillus* which is widespread in tropical Asia.

Corolla at maturity with clusters of stellate hairs on farinose background; bracts at least half as long or longer than ovary and calyx tube; calyx tube as long as the ovary; anthers c. 2 mm long ..1. *S. glauca*

Corolla at maturity lanate-pubescent; bracts broadened, approximately one half or less the length of ovary and calyx tube; calyx tube reduced to a rim less than 1 mm high; anthers c. 3 mm long ...2. *S. ovalis*

1. Septulina glauca (*Thunb.*) *v. Tieghem* in Bull. Soc. bot. Fr. 42: 263 (1895). Type: Cape, Saldanha Bay, *Thunberg* s.n. (UPS; see note below).

Loranthus glaucus Thunb., Prodr. 58 (1794); Fl. Cap. ed. Schultes 295 (1823); Eckl. & Zeyh., Enum. 358 (1837); Harv. in F.C. 2: 573 (1862); Sprague in F.C. 5, 2: 106 (1915); Mason, W. Cape Sandveld Flow. pl. 45: 1 (1972). *Dendrophthoe glauca* (Thunb.) Mart. in Flora 1: 109 (1830). *Scurrula glauca* (Thunb.) G. Don, Gen. Syst. 3: 424 (1834). *Taxillus glaucus* (Thunb.) Danser in Verh. K. Akad. Wet., sect. 2, 29, 6: 124 (1933). *Septulina glauca* var. *glauca*, Balle in Mitt. bot. StSamml., Münch. 7: 178 (1968); in F.S.W.A. 22: 9 (1969).

L. canescens Burch., Trav. 2: 90 (1824); DC., Prodr. 4: 304 (1830). *Scurrula canescens* (Burch.) G. Don, Gen. Syst. 3: 423 (1834). Type: Cape, Hanover, near Renoster Poort, *Burchell* 2119 (K, holo.!).

L. burchellii (DC.) Eckl. & Zeyh., Enum. 358 (1837), pro parte quoad spec. cit.

L. longitubulosus Engl. & Krause in Bot. Jb. 51: 455 (1914). Syntypes: South West Africa, Small Karas mountain, *Engler* 6662 (B); Kanus on Geiab River, *Dinter* 3071 (SAM!).

Small shrubs up to about 0,5 m high. *Leaves* sometimes fascicled, blades mostly oblanceolate-obovate, 10–15 × 3–7 mm, stellate-canescent when young, to scattered stellate-pubescent or glabrous with age; petioles 2–3 mm long. *Flowers* usually 2–3 in the axils. *Corolla* at anthesis mostly cylindrical, grey-green and dull reddish, lightly to moderately stellate-pubescent, 25–40 × 2–3 mm; buds stellate-canescent, cylindrical to slightly expanded at apex. *Calyx* reduced to a 5-toothed rim. *Berries* mostly ellipsoid, 7–8 mm long. *Flowering* March through November and possibly throughout the year; $n=9$. Fig. 15.

Western and interior Cape Province to south-western South West Africa. A frequent parasite on species of *Lycium*, but also reported on *Mesembryanthemum* and *Rhus* (Map 13).

Vouchers: *Dinter* 5215; 6185; *Hutchinson* 294; *Merxmüller & Giess* 2264.

No specimen of *Loranthus glaucus* is found in Thunberg's herbarium. Juel (1918) suggested that *L. incanus* was an alternative name used by Thunberg for this species and there are two specimens preserved of it. The specimen (No. 8806) might prove to be the type of *L. glaucus* as it was collected by Thunberg and agrees with the description by having a single flower (uniflorus) and ovate leaves.

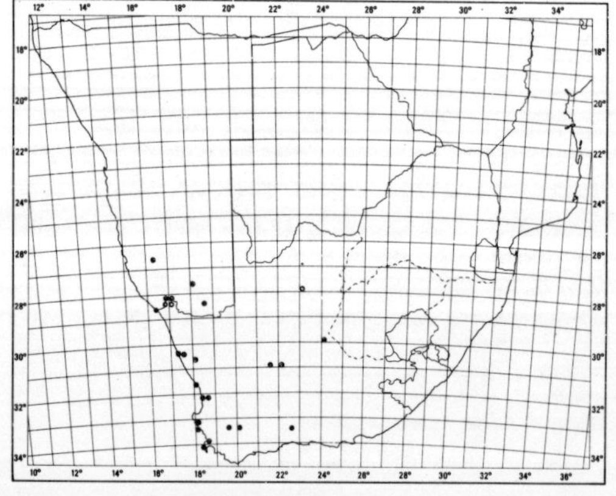

MAP 13.—●**Septulina glauca**
○S. ovalis

2. **Septulina ovalis** (*E. Mey. ex Harv.*) *v. Tieghem* in Bull. Soc. bot. Fr. 42: 263 (1895). Syntypes: Cape, Kaus, *Drège* s.n.; between Natvoet and Gariep, *Drège* s.n. (K!).

Loranthus ovalis E. Mey. ex Harv. in F.C. 2: 575 (1862); Sprague in F.C. 5, 2: 105 (1915). *Taxillus ovalis* (E. Mey. ex Harv.) Danser in Verh. K. Akad. Wet., sect. 2, 29, 6: 125 (1933). *Septulina glauca* var. *ovalis* (E. Mey. ex Harv.) Balle in Mitt. bot. StSamml., Münch. 7: 179 (1968); in F.S.W.A. 22: 10 (1969).

Shrubs to 1 m high, new growth densely stellate-tomentose. *Leaves* mostly obovate to oblong-elliptical, 15–30 × 10–20 mm, stellate pubescence often superimposed on a canescent background, sometimes glabrous with age; petioles 3–5 mm long. *Flowers* 1 or 2 in axils. *Corolla* at anthesis with somewhat swollen base, and often expanded apically, covered evenly with short grey-green stellate tomentum; buds lanate. *Calyx* 1–2 mm long, rim irregular. *Berries* ellipsoid, 11–12 mm long, red. *Flowering* from September to May, and possibly throughout the year. Fig. 15.

Extreme north-western Cape Province, the Kuruman region and south-western South West Africa. Parasitic mostly on *Tamarix* spp., but also occurring on *Lycium* (Map 13).

Vouchers: *Merxmüller & Giess* 28687, 3344; *Giess, Volk & Bleissner* 5404; *Giess* 13845.

This species is closely related to *S. glauca* but both the number and nature of the differences appear too great for conspecific classification as proposed by Balle. Although some characteristics approach each other there is little evidence that the species intergrade. Few collections of *S. ovalis* are available an d additional field studies are needed.

VISCACEAE

by D. WIENS* and H. R. TÖLKEN**

Shrubby or herbaceous, brittle, perennial, aerial hemiparasites of other dicotyledons or gymnosperms, glabrous to variously pubescent, often with swollen, articulated nodes. *Leaves* opposite, simple, entire, evergreen, sometimes reduced to scales, exstipulate. *Flowers* minute (c. 2 mm across) monochlamydeous, unisexual, solitary or clustered at the nodes, or in axillary spikes or dichasia. *Perianth segments* 2–4, valvate. *Staminate flowers* with stamens opposite to and as many as the perianth segments, episepalous or free; style vestigial or lacking. *Pollen* spherical, spined or smooth. *Pistillate flowers* with simple style and linear or capitate stigma. *Ovary* inferior, uniloculate, the ovules undifferentiated with embryo sacs originating within a short placental column. *Fruit* a 1-seeded berry or drupe with a viscous layer inside the vascular bundles. *Seeds* without testa; embryo mostly cylindrical, usually with 2 cotyledons. *Basic chromosome number* x=14.

A family of 7 genera and about 450 species, widely distributed through the tropical and north temperate regions of the world.

Dyer (Gen., 1975) included this family under Loranthaceae. For discussion of concepts and synonymy see Barlow (Proc. Linn. Soc. New South Wales 89: 268–272; 1964) and Kuijt (Brittonia 20: 136–147; 1968).

2093   **VISCUM**

Viscum L., Sp. Pl. ed. 1, 2: 1023 (1753); Gen. Pl. ed. 5: 448 (1754); Thunb., Fl. Cap. ed. Schultes 153 (1823); DC., Prodr. 4: 277 (1830); Harv. in F.C. 2: 578 (1862); Sprague in F.T.A. 6, 1: 393 (1910); in F.C. 5, 2: 121 (1915); Balle in F.S.W.A. 22: 11 (1969); Dyer, Gen. 1: 48 (1975). Type species: *V. album* L.

V. sect. *Aspidixia* Korth. in Verhandl. Batav. Genootsch. Kunsten 17: 235 (1839), as *Aspiduxia;* Benth. & Hook.f., Gen. Pl. 3, 1: 213 (1880), as *Aspiduxia;* Engl. in Pflanzenfam. ed. 1, 3, 1: 194 (1894); Sprague in F.C. 5, 2: 122 (1915); Engl. & Krause in Pflanzenfam. ed. 2, 16b: 202 (1935). *Aspidixia* (Korth.) v. Tieghem in Bull. Soc. bot. Fr. 43: 191 (1896). *V.* (sect. *Botryoviscum* Engl.) subsect. *Aspidixia* (Korth.) Engl. in Pflanzenfam. ed. 1, Nachtr. 1: 140 (1897).

V. sect. *Ploionixia* Korth. in Verhandl. Batav. Genootsch. Kunsten 17: 235 (1839), as *Ploionuxia;* Benth. & Hook.f., Gen. Pl. 3, 1: 213 (1880), as *Ploionuxia;* Engl. in Pflanzenfam. ed. 1, 3: 194 (1894), as *Pleionuxia;* Sprague in F.C. 5, 2: 122 (1915); Engl. & Krause in Pflanzenfam. ed. 2, 16b: 201 (1935). *V.* (sect. *Botryoviscum* Engl.) subsect. *Ploionixia* (Korth.) Engl. in Pflanzenfam. ed. 1, Nachtr. 1: 140 (1897).

Shrubby (rarely minute herbs), monoecious or dioecious, glabrous, hemiparasitic, aerial parasites on dicotyledons (rarely gymnosperms). *Branching* usually dense and intricate, forked or whorled. *Internodes* rounded or compressed, sometimes ribbed and twisted 90 degrees forming a decussate leaf and branching pattern. *Inflorescence* a typical or modified dichasium subtended by a pair of usually fused bracts (bracteal cup). *Dichasia* sessile or peduncled, solitary or fascicled, axillary, or axillary and terminal. *Monoecious plants* with central flower usually staminate and lateral ones pistillate, or occasionally with all flowers of the dichasium

* Department of Biology, University of Utah, Salt Lake City, Utah 84112, U.S.A.
    This work was completed with the assistance of NSF Grant DEB 75-21170 and while a Research Fellow at the Botanical Research Institute, Pretoria.
** Formerly Botanical Research Institute, Pretoria. Present address: State Herbarium, North Terrace, Adelaide, South Australia 5000, Australia.

Published: 1979.

staminate or pistillate. *Dioecious plants* with staminate dichasia usually bearing 3(2) flowers. *Pistillate dichasia* with a solitary flower in the bracteal cup. *Staminate flowers* 3–4-merous, anthers dehiscing by numerous pores. *Pistillate flowers* 3–4-merous. *Berries* white, yellow, orange, or red, smooth or warty, pedicelled or sessile in the bracteal cup; perianth segments occasionally persistent; style usually persistent.

A genus of approximately 100 species, widely distributed in the African (including Madagascar) and Asian tropics, with significant extensions into the north temperate zones of Europe and Asia; in the south temperate regions only in Southern Africa.

1 Shrub 0,25 m to 1 m or higher, woody at least at the base; parasitic on diverse dicotyledons (rarely on *Aloe*):

  2 Plant leafless or with scale leaves only:

   3 Stems conspicuously flattened, especially when young, older stems becoming rounded:

    4 Internodes mostly 30–35 × 4–6 mm, margin not distinctive; berries sessile, ellipsoid, 5–6 mm long, uniformly warty when young, nearly smooth at maturity; style about as long as stigma ....1. *V. combreticola*

    4 Internodes mostly 15–20 × 2–4 mm, with pale-yellow margin, which forms a narrow wing on older stems; berries with pedicels 1 mm long at maturity, ovoid-ellipsoid, 4–5 mm long, the upper half mostly warty; style persistent, about twice as long as stigma .... 2. *V. anceps*

   3 Stems rounded at all developmental stages:

    5 Berries heavily warted when young, less so at maturity:

     6 Berries sessile at maturity, when young subcylindrical, truncate apically, slightly constricted below the apical rim; internodes 40–60 mm long; staminate dichasia always 3-flowered, in bracteal cups; parasitic on *Ficus* spp. ....3. *V. menyharthii*

     6 Berries with a pedicel 1–2 mm long, when young obovate, not truncate or constricted below the apical rim; internodes 15–25 mm long; staminate flowers 1, 2, or 3, not always in bracteal cups; parasitic mostly on *Acacia* spp. ....4. *V. verrucosum*

    5 Berries smooth at all developmental stages:

     7 Berries sessile, white; nodes with minute but conspicuous scale leaves (c. 1 mm long) sometimes protruding at nearly right angles; primary and secondary branches approximately equal in width; bracteal cups the same colour as the stem ....5. *V. capense*

     7 Berries with a stout pedicel c. 2 mm long, pale yellow; scale leaves not especially evident; primary stems often twice as thick as the laterals; bracteal cups with minute but distinctive white margins ....6. *V. continuum*

  2 Plant with well developed leaves:

   8 Dichasia with central flower staminate and lateral flowers pistillate, or the 3 flowers all pistillate, or less commonly all staminate; at least some of the bracteal cups bearing 2 (rarely 3) berries:

    9 Berries warty, especially when young, less so when mature:

     10 Internodes of leafy stems essentially smooth (wrinkled when dry), mostly 8–15 mm long; leaves 12–17 × 8–12 mm; berries 3–4 mm long; lowlands of Natal and eastern Cape ....7. *V. obovatum*

     10 Internodes of leafy stems 6-ribbed, mostly 15–20 mm long; leaves 20–30 × 12–20 mm; berries 5–6 mm long; highlands of the central and northern Transvaal ....8. *V. spragueanum*

    9 Berries smooth at all developmental stages:

     11 Berries sessile-subsessile in the bracteal cup, white at maturity; leaves yellow-green, thin, with usually undulate margin, apex rounded-obtuse, distinctly petiolate, but merging gradually into the cuneate base ....9. *V. nervosum*

     11 Berries pedicellate, emerging at least 1,5 mm above bracteal cup, light yellow to orange at maturity; leaves often glaucous, coriaceous, margins even and smooth, some leaves minutely apiculate, sessile-subsessile, base not cuneate:

      12 Leaves 6–12 × 3–8 mm, base rounded-obtuse (if acute then blade usually 5 mm or less wide):

       13 Berries pale yellow, c. 3 mm long, with short (1,5 mm long) pedicel; leaves usually less than 8 × 4 mm; internodes usually shorter than 12 mm; northern Cape Province to central South West Africa ....10. *V. schaeferi*

       13 Berries orange, c. 4–5 mm long, with pedicel 3–4 mm long; leaves usually more than 8 × 4 mm; internodes usually longer than 12 mm; widespread throughout Southern Africa ....11. *V. rotnudifolium*

    12 Leaves 15–25 × 8–12 mm, base mostly acute ....12. *V. pauciflorum*

8 Dichasia with staminate flowers only, or the flowers (or berries) all solitary and pistillate in the bracteal cups:

    14 Berries warty when young, but nearly smooth at maturity; leaves often minutely serrulate ....13. *V. subserratum*

    14 Berries always smooth at maturity; leaves entire:

      15 Leaves obovate-orbicular, 6–12 × 5–8 mm; stems rounded (both leaves and stems highly succulent when fresh, becoming wrinkled and distorted when dry); berries sessile-subsessile ....14. *V. crassulae*

      15 Leaves oblanceolate-ovate, 15–50 × 8–20 mm; young stems 6-ribbed (not especially succulent); berries with pedicel 3–6 mm long at maturity:

        16 Berries white at maturity, 6–7 mm long, often 3–6 per node; pedicels 5–6 mm long, expanding upward to nearly twice the basal width; the persistent style on mature fruit c. 1,0 mm long ....15. *V. obscurum*

        16 Berries orange at maturity, 4–5 mm long, usually 2 per node; pedicels 3–4 mm long, not conspicuously enlarged upward; the persistent style on mature fruit c. 0,5 mm long ....16. *V. oreophilum*

1 Minute herb only several mm high; parasitic on succulent *Euphorbia* spp. ....17. *V. minimum*

1. Viscum combreticola *Engl.* in Bot. Jb. 40: 542 (1908); Sprague in F.T.A. 6, 1: 404 (1911); in F.C. 5, 2: 130 (1915); Burtt Davy, Fl. Transv. 467 (1932); Balle in F.C.B. 1: 380 (1948); Letty, Wild Flow. Transv. pl. 61, 2 (1962). Type: Transvaal, Magaliesberg at Buffelspoort, *Engler* 2840a (B†; K!).

V. dichotomum sensu Harv. in F.C. 2: 581 (1862), pro parte, excl. syn.

Leafless, dioecious shrubs of relatively large size, 1–2 m high, often becoming pendulous with age, typically yellowish green; younger branches strongly flattened and ribbed; older branches rounded; basal internodes of younger branches (20–) 30–35 (–40) × 4–6 mm. *Staminate flowers* in dichasia (occasionally 2-flowered), solitary in axils of scale leaves on younger branches, often in fascicles of 4–8 on older stems. *Pistillate flowers* solitary in the bracteal cups, borne singly in axils of scale leaves of younger branches, bracteal cups rounded and bilobed. *Berries* ellipsoid, 6–7 mm long, sessile, warty when young but nearly smooth and orange at maturity, apex truncated; style persistent. *Flowering* inconsistent, but possibly mostly from February–May, and perhaps scattered throughout the year; *n*=14. Fig. 16.

Parasitic on species of *Acacia, Combretum, Croton, Diplorhynchus, Dombeya, Heteropyxis, Maytenus, Melia, Strychnos, Terminalia* and *Vangueria*, but primarily on *Combretum* spp. in Transvaal and northern Orange Free State; also in Rhodesia (Map 14).

Vouchers: *Codd* 2607; *Galpin* 9075; 11627; *Marloth* 3805.

A well defined species, but similar to *V. anceps* vegetatively. Harvey united these two species under *V. dichotomum*, an Asian species, which they both resemble superficially.

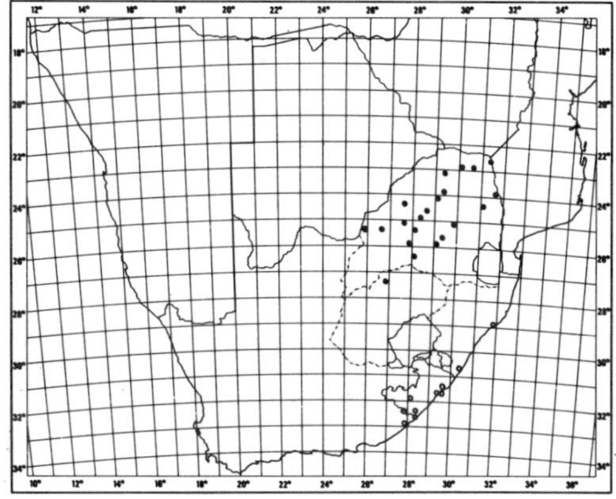

MAP 14.—●**Viscum combreticola**
○**V. anceps**

2. Viscum anceps *E. Mey. ex Sprague* in F.T.A. 6, 3: 407 (1911); in F.C. 5, 2: 130 (1915); Batten & Bokelmann, Wild Flow. E. Cape Prov. pl. 54, 5 (1966); Ross, Fl. Natal 153 (1973). Type: Cape, between Umtata and St John's River, *Drège* s.n. (K, lecto.!).

Aspidixia anceps E. Mey. ex v. Tieghem in Bull. Soc. bot. Fr. 43: 193 (1896), nom. nud.

V. dichotomum sensu Harv. in F.C. 2: 581 (1862), pro parte, excl. syn.

Leafless, dioecious shrubs of moderate size, mostly 0,5–1 m high, often pale yellowish green; younger branches strongly flattened, minutely ribbed; older stems rounded; basal internodes of younger branches (10–) 15–20 (–30) × 3–4 mm, bearing a minute but distinct, pale yellow margin, this transformed into a wing on older stems. *Staminate flowers* in dichasia (sometimes 2-flowered), sessile, mostly solitary at nodes of younger branches. *Pistillate flowers* solitary in bracteal cup, sessile at nodes of younger branches; bracteal cup acute, bilobed, margin scarious. *Berries* ovoid-ellipsoid, 4–5 mm long, developing a short pedicel (c. 1 mm long), strongly warty only on upper half at maturity, dull orange. *Flowering* approximately June through July; $n=14$. Fig. 16.

Parasitic on species of *Acacia, Calpurnia, Citrus* and *Fagara,* from central Natal through Transkei to eastern Cape Province (Map 14).

Vouchers: *Flanagan* 197; *Galpin* 3421; *Pegler* 1809.

A relatively distinct species, perhaps most closely related to *V. combreticola* and *V. shirense* Sprague (a Malawian species). Its relationship to *V. junodii* auct. from Mozambique needs further study.

3. Viscum menyharthii *Engl. & Schinz* in Denkschr. Akad. Wiss. Wien, Math.-Naturw. Kl. 78: 410 (1906), as *menyhartii.* Type: Mozambique, Boruma, *Menyharth* s.n. (K!).

Leafless, dioecious shrubs of moderate size, 0,5–1 m high, occasionally pendulous. *Stems* rounded, often greatly elongated; basal internodes of younger branches (30–) 40–60 ×2–3 mm. *Staminate flowers* in dichasia, mostly (2–) 4–6 per node, but up to 10 on terminal nodes. *Pistillate flowers* solitary in bracteal cups, 2–4 per node. *Berries,* when young, cylindrical-subovoid, warty, truncate slightly constricted below the rim, at maturity ovoid, 5–6 mm long, faintly tuberculate to smooth, sessile, remaining truncate, orange; style persistent. *Flowering* probably mainly in September and October; $n=14$. Fig. 16.

Parasitic only on *Ficus* spp. in the northern and north-eastern Transvaal; apparently disjunct in South West Africa (Map 15).

Vouchers: *Codd & De Winter* 5561; *Wiens* 5315.

A highly distinct species with a primarily tropical distribution, reaching only the northern fringes of Southern Africa. The fragmentary material of *V. rigidum* Engl. & Krause we have seen suggests this is only a form of *V. menyharthii.*

4. Viscum verrucosum *Harv.* in F.C. 2: 581 (1862); Sprague in F.T.A. 6, 3: 408 (1911); in F.C. 5, 2: 134 (1915); Burtt Davy, Fl. Transv. 467 (1932); Ross, Fl. Natal 153 (1972); Gibson, Wild Flow. Natal pl. 30,8 (1975). Syntypes: Transvaal, Magaliesberg, *Sanderson* s.n. (K!); Natal, Mooi River Valley, *Sutherland* s.n. (K!; PRE!).

Leafless, dioecious shrubs of moderate size, mostly 0,5–1 m high, usually densely and intricately branched, mostly pale, light green. *Stems* rounded; basal internodes of younger branches 15–25×1–2 mm. *Staminate flowers* in dichasia (sometimes 2-flowered) subtended by the typical bracteal cup, but staminate flowers also borne adventitiously and singly, or in groups of 2 or 3 at nodes outside bracteal cups, often resulting in fascicles of up to 8 flowers. *Pistillate flowers* borne solitary either in bracteal cups or sessile at nodes, mostly 2 per node, occasionally 3–4. *Berries* at maturity rounded, 5–6 mm high, faintly warty or nearly smooth, pale yellow-orange, developing a pedicel 1–2 mm long, immature berries obovate-elliptic, densely warty; style persistent. *Flowering* March through July; $n=14$. Fig. 17.

Parasitic mainly on species of *Acacia,* but also on *Combretum:* occurring in Botswana, Transvaal, Swaziland and Natal; apparently disjunct in central South West Africa (Map 15).

Vouchers: *Galpin* 13335; *Leistner* 3201; *Strey* 3683, 9823.

A distinct and widely distributed species. The South West African collections (*Giess* 10986; *Merxmüller & Giess* 1604) are somewhat different from those in the rest of Southern Africa and field studies should be made to determine their status.

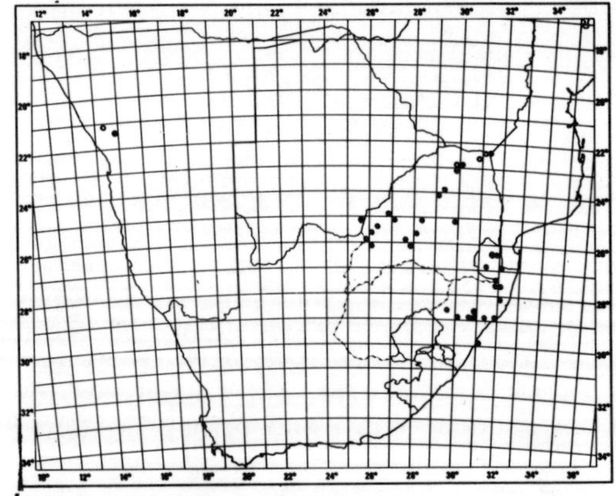

MAP 15.—○Viscum menyharthii
●V. verrucosum

5. Viscum capense *L.f.,* Suppl. 426 (1781); Thunb., Fl. Cap. ed. Schultes 154 (1823); Harv. in F.C. 2: 581 (1862); Sprague in F.C. 5, 2: 132 (1915). Type: Cape, *Sparrman* in LINN 1166.9 (LINN, holo.; PRE, micro-fiche!).

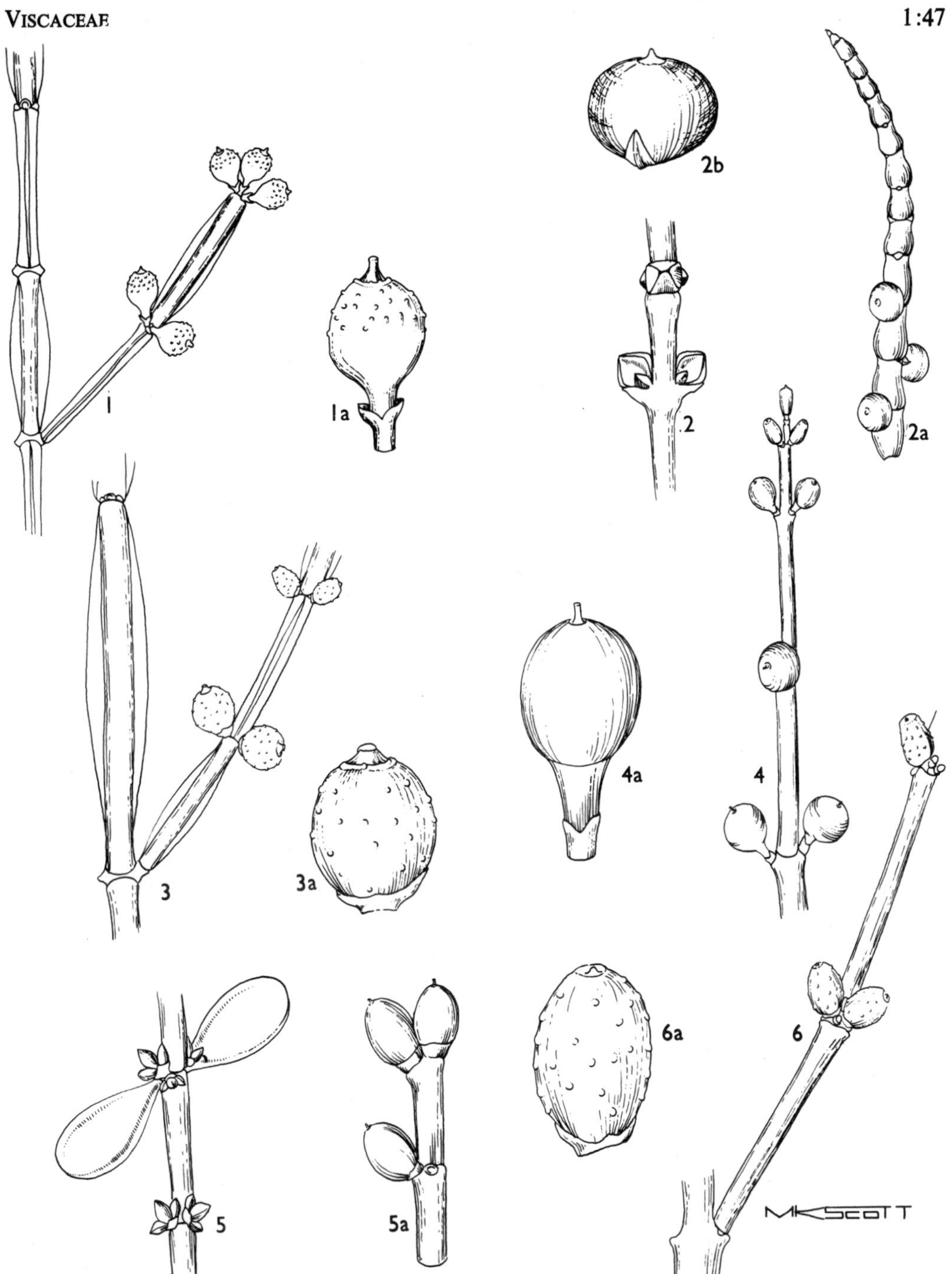

Fig. 16.—1, **Viscum anceps**, fruiting twig, ×1; 1a, fruit, ×3 (*Wiens* 5364). 2, **V. capense** subsp. **capense**, flowering twig, ×2 (*Wiens* 5416); 2a, fruiting twig, ×1; 2b, fruit, ×3 (*Wiens* 5410). 3, **V. combreticola**, fruiting twig, ×1; 3a, fruit, ×3 (*Lang* in TRV 32074). 4, **V. continuum**, fruiting twig, ×1; 4a, fruit, ×3 (*Wiens* 5397). 5, **V. crassulae**, flowering twig, ×1; 5a, fruiting twig, ×1 (*Wiens* 5389). 6, **V. menyharthii**, fruiting twig, ×1; 6a, fruit, ×3 (*Codd & De Winter* 5561).

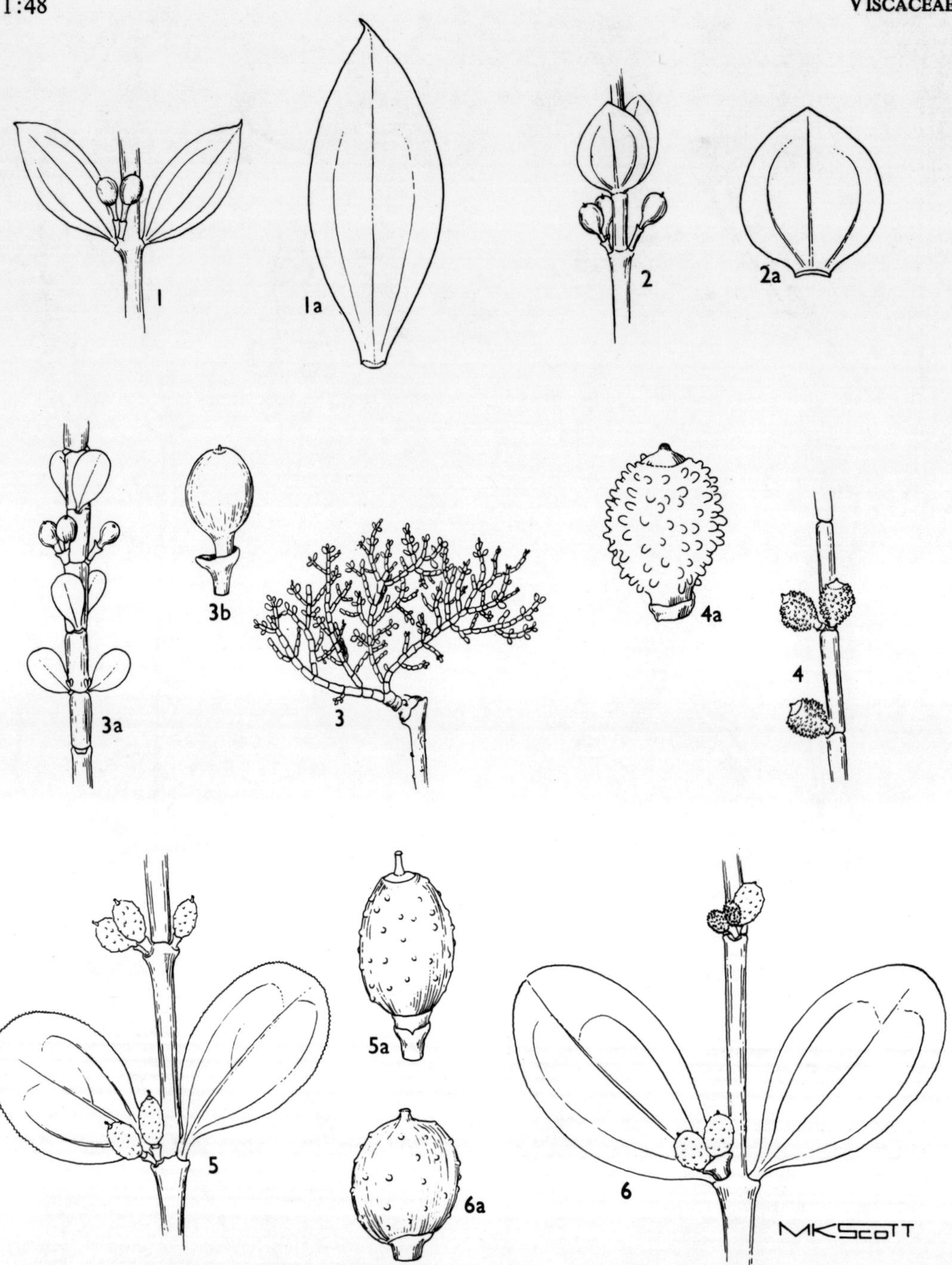

FIG. 17.—1, **Viscum pauciflorum,** fruiting twig, ×1; 1a, leaf, ×2 (*Wiens* 5411). 2, **V. rotundifolium,** fruiting twig, ×1; 2a, leaf, ×2 (*Repton* 1662). 3, **V. schaeferi,** habit of plant, ×0,25; 3a, fruiting twig, ×1; 3b, fruit, ×3 (*Wiens* 5420). 4, **V. verrucosum,** fruiting twig, ×1; 4a, fruit, ×3 (*Rodin* 4218). 5, **V. subserratum,** fruiting twig, ×1; 5a, fruit, ×3 (*Thorncroft* 2227). 6, **V. spragueanum,** fruiting twig, ×1; 6a, fruit, ×3 (*Jacobsen* 2048).

Leafless or with scale-like leaf rudiments, monoecious or dioecious semi-shrubs mostly less than 0,5 m high, often forming dense rounded, or occasionally elongate, clusters, often glaucous. *Stems* rounded; basal internodes of younger branches (6–) 8–12 (–15)×2–3 mm, younger shoots often about the same diameter as the primary shoots. *Leaf scales* acute, c. 1 mm long, often projecting conspicuously at right angles from stem. *Staminate flowers* solitary in the bracteal cup, often 3–5-fascicled at the nodes. *Pistillate flowers* borne similarly. *Berries* rounded, 3–4 mm long, sessile, smooth, white (almost translucent when fresh and the mesocarp watery, not viscid). *Flowering* from approximately July through October; $n=10$.

Parasitic on species of *Chrysanthemoides*, *Erica*, *Euclea*, *Maytenus*, *Passerina*, *Pterocelastrus*, *Rhus*, *Scolopia*, and *Scutia;* from the south-western Orange Free State, southern and western Cape Province to central South West Africa; disjunct in the central Transvaal.

This species is here interpreted as consisting of two subspecies segragated solely on the basis of monoecy or dioecy, although other characters might reinforce this subdivision when additional information is available. Dioecy and monoecy are remarkably species-constant characters in all African, Madagascan, Asian, and Australian species thus far studied. The two subspecies here recognized are designed to call attention to this biological feature, since the characters are constant for the rather broad geographical regions indicated. The entire complex is in need of an extensive, critical study. For the present, no further taxonomic separation appears warranted.

Plants dioecious ....(a) subsp. *capense*

Plants monoecious ....(b) subsp. *hoolei*

(a) subsp. **capense.**

Viscum capense L.f., Suppl. 426 (1781); Thunb., Fl. Cap. ed. Schultes 154 (1823); DC., Prodr. 4: 283 (1830); Harv. in F.C. 2: 581 (1862), pro parte excl. *V. continuum* E. Mey. et *Zeyher* 749 partim; Sprague in F.C. 5, 2: 132 (1915); Adamson, Fl. Cape Penins. 343 (1950); Kidd, Wild Flow. Cape Penins. pl. 39, 2 (1950); Balle in Mitt. bot. StSamml., Münch. 7: 190 (1968); in F.S.W.A. 22: 12 (1968). *Aspidixia capensis* (L.f.) v. Tieghem in Bull. Soc. bot. Fr. 43: 193 (1896).

V. robustum Eckl. & Zeyh., Enum. 358 (1837). *Aspidixia robusta* (Eckl. & Zeyh.) v. Tieghem in Bull. Soc. bot. Fr. 43: 193 (1896). Type: Cape, Namaqualand, *Ecklon & Zeyher* 2279 (K!; SAM!).

V. rigidum Engl. & Krause in Bot. Jb. 51: 471 (1914); Sprague in F.C. 5, 2: 134 (1934). Type: South West Africa, Gross Karas am Us-Rivier, *Engler* 6446 (B†).

V. dielsianum Dinter ex Neusser in Mitt. bot. StSamml., Münch. 1: 344 (1953). Type: South West Africa, Witpütz, *Dinter* s.n.

On various hosts from the Cape Peninsula northward through western Cape Province to central South West Africa, but eastward along the coast only to the vicinity of Swellendam; the disjunct populations in the central Transvaal are also dioecious and are placed under the typical subspecies. Fig. 16; Map 16.

Vouchers: *Wiens* 5409; 5416; 5435.

The populations of the Cape Peninsula are a deep green, whereas some of the maritime populations east of the Cape are yellowish with characteristically short (less than 10 mm) stout internodes rather consistent in size. The populations from the western Cape to South West Africa are glaucous, brownish, and somewhat similar to those of the central Transvaal, although the Transvaal populations flower at least a month or more (July–August) before those of the western Cape.

The dull yellowish colour and the short stout internodes of *V. robustum* suggest a possible pathological or teratological condition, and any resemblance to the maritime populations east of the Cape Peninsula is fortuitous. Little evidence exists to support Sprague's acceptance of this species.

V. dielsianum does not appear to differ in any significant way from other populations of this species in South West Africa.

No type specimen of *V. rigidum* Engl. & Krause could be traced. *V. capense* is the only species which occurs in the area and it has scale-like leaf rudiments, short fleshy internodes and sizable bracts (c. 2 mm long). The observation by Engler and Krause that *V. rigidum* is closely related to *V. menyharthii* is misleading and has probably been based solely on the slightly warted fruits described, which, according to the authors, are very immature.

(b) subsp. **hoolei** *Wiens* in Bothalia 12: 423 (1978). Type: Cape, Slaaikraal Farm, 8 km north-west of Grahamstown, *Wiens & Hoole* 5385f (K, holo.!; PRE!; UT!).

On various hosts eastward from the Swellendam region through the eastern Cape Province to the south-western Orange Free State (Map 16).

Vouchers: *Wiens* 5391; 5402.

Although this subspecies is monoecious, the distribution of pistillate and staminate flowers is not generally equal, and there is usually a predominance of one sex on flowering shoots, making the monoecious condition sometimes difficult to detect. Whereas the internodes of typical *V. capense* are usually shorter than 10 mm, stout, and of rather uniform dimensions (especially the population from the south-western Cape), those of subsp. *hoolei* are often longer than 10 mm, and the terminal and lateral shoots tend to be thinner than the primary shoots. The consistency of these stem characters, however, needs further study.

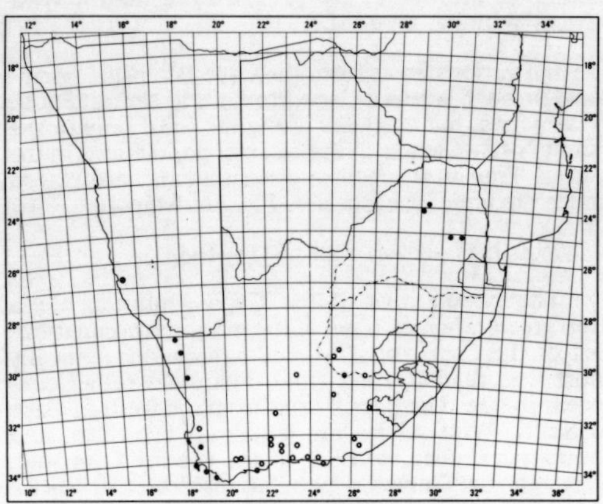

MAP 16.—●Viscum capense subsp. **capense**
　　　　　○Viscum capense subsp. **hoolei**

6. Viscum continuum *E. Mey. ex Sprague*

in F.T.A. 6, 1: 410 (1911); in F.C. 5, 2: 133 (1915); Batten & Bokelmann, Wild Flow. E. Cape Prov. pl. 54, 6 (1966). Type: Cape, between Driekoppen and Bloedrivier, *Drège* s.n. (K, lecto.!).

V. capense sensu Harv. in F.C. 2: 581 (1862), pro parte, quoad *Drège* s.n. and *Burke* s.n.

Leafless, dioecious shrubs of relatively large size, 1 m or higher, often becoming pendulous, somewhat light green. *Stems* rounded; basal internodes of younger branches (10–) 15–25 ×1 (–2) mm; younger shoots usually half as wide as primary stems. *Leaf scales* not projecting prominently from stems, margins whitish. *Staminate flowers* in dichasia, or more commonly borne singly or doubly at nodes without bracteal cups. *Pistillate flowers* solitary in bracteal cups, these borne singly in axils of leaf scales. *Berries* ovoid-ellipsoid, 4–5 mm high, smooth, pale yellow, at maturity developing a long, stout pedicel c. 2 mm long, nearly as wide as long; style persistent. *Flowering* in July and August; $n=14$. Fig. 16.

Parasitic only on *Acacia* spp. and endemic to the Little Karroo and adjoining regions as far as the eastern Cape (Map 17).

Vouchers: *Acocks* 11905; *Burtt Davy* 12274; *Marloth* 11261.

A distinctive species not especially closely related to *V. capense* as believed by Harvey. It is possibly more closely related to *V. verrucosum*, as suggested by Sprague.

7. Viscum obovatum *Harv.* in F.C. 2:

579 (1862); Sprague in F.C. 5, 2: 122 (1915); Ross, Fl. Natal 153 (1972). Type: Natal, Durban, *Gerrard & McKen* 659 (TCD, holo.!).

V. pulchellum Sprague in Kew Bull. 1915: 81 (1915); in F.C. 5, 2: 123 (1915). Type: Natal, Tugela River, *Gerrard* 1649 (K, holo.!; PRE!).

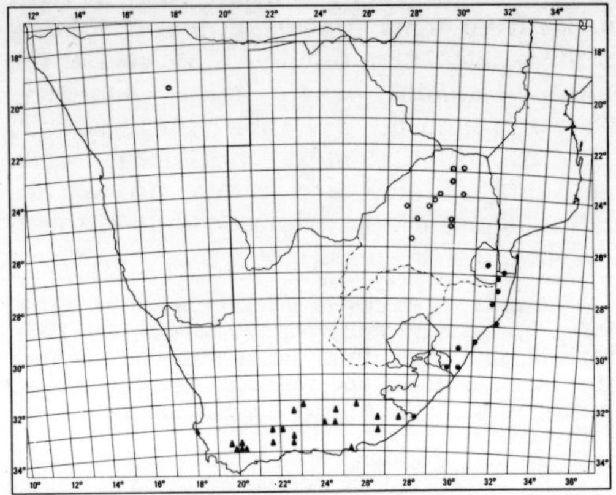

MAP 17.—▲Viscum continuum
　　　　　●V. obovatum
　　　　　○V. spragueanum

Leafy, monoecious shrubs of small to moderate size, mostly less than 0,5 m high, often forming dense, rounded clusters. *Stems* rounded, often with swollen nodes and shortened internodes; basal internodes of young branches 8–15 (–20) ×2 mm. *Leaves* mostly obovate-rounded, 12–17 ×8–12 mm, rounded apically, cuneate into the base; petiole 2–3 mm long. *Dichasia* with central flower mostly staminate, lateral pistillate, or with all flowers the same sex. *Berries* elliptic, 3–4 mm high, sessile, warty when young, less so at maturity, truncate apically, dull orange; style persistent. *Flowering* in March and April, but possibly continuously or sporadically throughout the year; $n=12$. Fig. 18.

Parasitic on species of *Dichrostachys, Galpinia, Maytenus, Mimusops* and *Ochna*; from the coastal regions of the eastern Cape and Natal, to the Lebombo Mountains of Swaziland (Map 17).

Vouchers: *Strey* 6678; 9732; 10994.

Viscum pulchellum Sprague is placed in synonymy under this species. Sprague believed the species to be dioecious, but no further collections of it have apparently been made. The type specimen (a pistillate plant) has the flowers borne only singly in the bracteal cups, a condition not observed in *V. obovatum* which usually has at least 2 flowers per dichasium in some of the bracteal cups as is typical of monoecious *Visca*. The fruit and habit of the type of *V. pulchellum* are generally similar to *V. obovatum*, except the leaves are somewhat smaller in the former. In *V. obovatum*, however, the plants often tend to have one sex predominating, and it is possible the type specimen of *V. pulchellum* represented a portion of a plant which by chance was totally pistillate with only solitary flowers. Until further evidence is available it seems best to retain *V. pulchellum* under *V. obovatum*. The sexual distribution of flowers in this species should be studied further.

8. **Viscum spragueanum** *Burtt Davy*, Fl. Transv. 2: 466 (1932); Balle in Mitt. bot. StSamml., Münch. 7: 197 (1968); in F.S.W.A. 22: 13 (1968); Ross, Fl. Natal 153 (1972). Type: Transvaal, Moorddrif, *Leendertz* 2236 (K, holo.!; PRE!).

V. tarchonanthum Welw. ex v. Tieghem in Bull. Soc. bot. Fr. 43: 190 (1896), nom. nud.

Leafy, monoecious shrubs of moderate size mostly 0,5–1 m high, mostly dark green, densely and intricately branched; younger branches somewhat flattened and 6-ribbed, the rib below the leaves sometimes transformed into a wing; older branches rounded; basal internodes of younger branches mostly 15–20 × 2–3 mm, somewhat dilated at the nodes. *Leaves* mostly obovate-oblong to orbicular, 20–30 × 12–20 mm, apically rounded, often cuneate into base, 3(5)-nerved from base (sometimes only faintly); petiole subsessile to 3 mm long. *Dichasia* with central flower typically staminate, lateral pistillate developing a short, stout peduncle c. 1 mm long. *Berries* subcylindrical, densely warted, truncate when young, at maturity nearly smooth, ellipsoid to globose, 5–6 mm long, pale yellow-orange; style persistent, cylindrical; stigma approximately the diameter of the style. *Flowering* April through June (possibly longer); $n=23$. Fig. 17.

Parasitic on species of *Acacia, Combretum, Grewia, Rhus,* and *Vitex* in the central and northern Transvaal and in South West Africa (Map 17).

Vouchers: *Acocks & Hafström* 407; *Fries, Nordlindh & Weimarck* 1979, 1980; *Stauffer & Scheepers* 5257.

A distinct species, but frequently confused with *V. subserratum* with which it is sympatric. It is most easily distinguished from *V. subserratum* by the monoecious state which consistently results in at least 2 berries being produced in some bracteal cups. The berries are also distinctive at maturity, those of *V. spragueanum* are sessile with a persistent linear stigma, whereas *V. subserratum* has berries with short pedicels and persistent capitate stigmas. The leaves of *V. subserratum* are also often (but not consistently) minutely, dark serrulate on the apical half.

This is the only polyploid species of *Viscum* in the Southern African flora. The odd chromosome number suggests a dibasic origin and *V. subserratum* ($n=11$) might be one of its progenitors, thus explaining the resemblance to that species. The other possible progenitor might be *V. obovatum* ($n=12$).

The species apparently is disjunct in central South West Africa. The material is reasonably close to that from the Transvaal, but field studies should be conducted to determine the validity of maintaining the South West African populations in *V. spragueanum*.

9. **Viscum nervosum** *Hochst. ex A. Rich.*, Tent. Fl. Abyss. 1: 338 (1847); Sprague in F.T.A. 6, 1: 397 (1911); in F.C. 5, 2: 124 (1915); Balle in F.C.B. 1: 376 (1948); Ross, Fl. Natal 153 (1972). Syntypes: Abyssinia, near Worrhey, *Schimper* 678; Schahagenni, *Schimper* 211 (K!); sine loc., *Quarton Dillon*.

Leafy, monoecious shrubs of relatively small size, usually no more than 0,5 m high, often forming rounded, yellow-green masses with elongated, somewhat delicate branchlets. *Stems* rounded; younger branches 6-ribbed and somewhat flattened; older branches rounded; basal internodes of younger branches (10–) 15–20 (–25) × 1,0–1,5 mm; nodes often dilated. *Leaves* obovate-elliptic, 18–25 (–30) × 8–13 mm, mostly rounded-obtuse apically, cuneate into the base, conspicuously 3(–5)-nerved from the base, yellowish green, relatively thin, not as coriaceous as many *Visca*, margin often undulate or crisped; petiole rather indistinct when merging into blade, c. 3 mm long. *Dichasia* with central flower mostly staminate, lateral pistillate, or with all flowers the same sex, solitary in axils with a narrow elongate peduncle 2–3 mm long, 0,5 mm or less wide. *Berries* ellipsoid-globose, 4–5 mm long, essentially sessile, smooth, white; sepals and style often persistent. *Flowering* in March (possibly sporadically through much of the year); $n=14$. Fig. 18.

Parasitic with certainty only on species of *Rapanea* and *Syzygium* but probably also on other hosts from the Soutpansberg in Transvaal to Transkei (Map 18).

Vouchers: *Acocks* 11640; *Galpin* 2886.

This species is apparently widely distributed in eastern Africa as far north as Eritrea. The type is reasonably similar to our material and that from Kenya. Most *Visca*, however, do not have such extensive ranges and more detailed studies of this taxon should be made over its entire range.

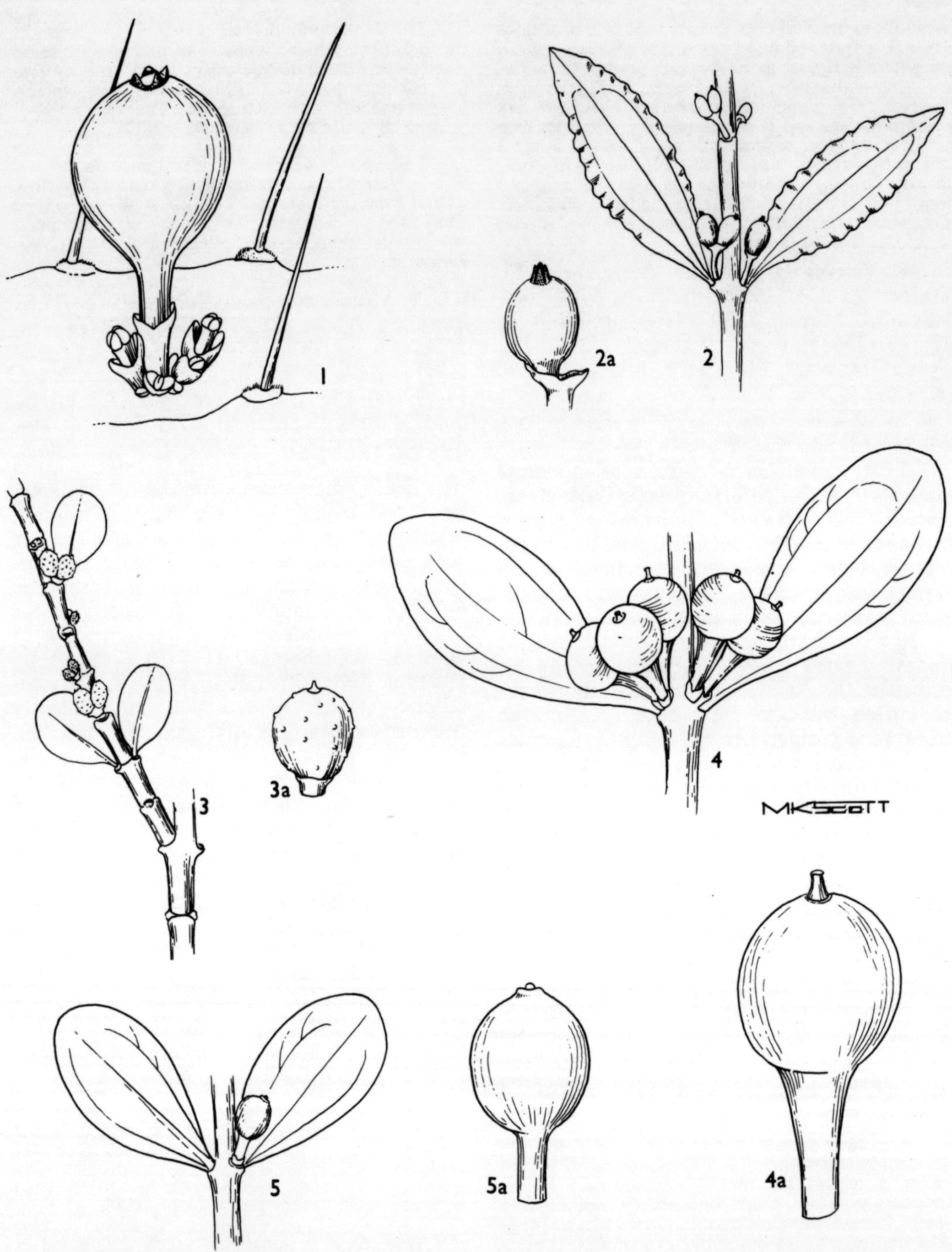

FIG. 18.—1, **Viscum minimum,** habit of plant growing on *Euphorbia* sp., ×3 (*Wiens* 5375). 2, **V. nervosum,** fruiting twig, ×1 (*Compton* 31680); 2a, fruit, ×3 (*Wiens* 5262). 3, **V. obovatum,** fruiting twig, ×1; 3a, fruit, ×3 (*Strey* 10994). 4, **V. obscurum,** fruiting twig, ×1; 4a, fruit, ×3 (*Wiens* 5356). 5, **V. oreophilum,** fruiting twig, ×1; 5a, fruit, ×3 (*Compton* 27859).

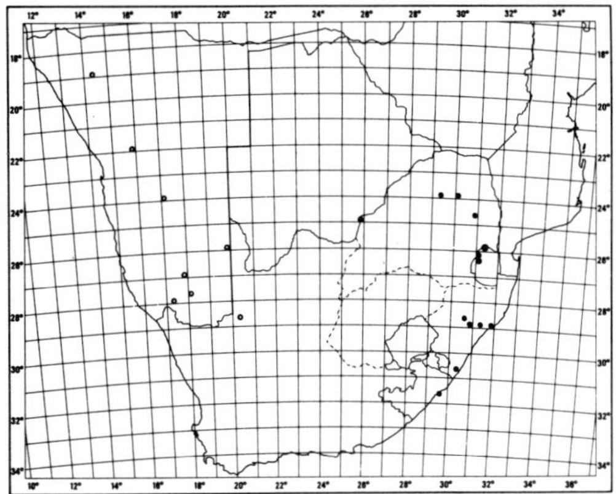

MAP 18.—●Viscum nervosum
○V. schaeferi

10. **Viscum schaeferi** *Engl. & Krause* in Bot. Jb. 51: 470 (1914); Sprague in F.C. 5, 2: 129 (1915). Syntypes: South West Africa, near Seeheim, *Schäfer* 465 (B †); *Engler* 6601 (B †).

V. rotundifolium sensu Balle in Mitt. bot. StSamml., Münch. 7: 193 (1968), pro parte; in F.S.W.A. 22: 12 (1968), pro parte, quoad syn. *V. schaeferi.*

Leafy, monoecious shrubs of small size, probably not higher than 0,2 m, stout, densely branched with short internodes, greyish green with small leaves, much resembling *Boscia*, one of its principal hosts. *Stems* rounded to slightly angled; basal internodes of younger branches 8–12 × 1–2 mm (rather consistent in size), older nodes slightly swollen. *Leaves* mostly oblong to oblanceolate, 6–8 × 3–4 mm, apex mostly obtuse, often minutely apiculate, tapering slightly into the sessile base. *Dichasia* with central flower typically staminate, lateral pistillate, usually 2 per axil. *Berries* ellipsoid, c. 3 mm long, smooth, forming a pedicel at maturity up to 1,5 mm long; style persistent. *Flowering* period unknown, possibly August–September. Fig. 17.

Parasitic on species of *Albizia, Boscia, Euclea,* and *Pappea* from South West Africa, northern Cape and western Transvaal (Map 18).

Vouchers: *Boss* in TRV 35870; 36051; *Merxmüller & Giess* 28147: *Kinges* 3171-2.

This species was accorded uncertain status by Sprague in F.C. 5, 2: 129 (1915) and considered synonymous with *V. rotundifolium* by Balle in F.S.W.A. 22: 13 (1968). Field studies show that this species is clearly distinct from *V. rotundifolium* by its

smaller, robust stature with rigid much branched habit, usually smaller leaves, and especially the smaller (up to 3 mm long) pale yellow berries. When parasitic on *Boscia* (and perhaps also other genera) this mistletoe (particularly the leaves) resembles its host to such an extent that mimicry might be suspected [See Barlow and Wiens in *Evolution* 31(1): 69–84; 1977].

11. **Viscum rotundifolium** *L.f.*, Suppl. 426 (1781); Thunb., Fl. Cap. ed. Schultes 154 (1823); DC., Prodr. 4: 279 (1830); Eckl. & Zeyh., Enum. 357 (1837); Harv. in F.C. 2: 580 (1862); Schinz in Bull. herb. Boissier sér. 1, 4, App. 3: 55 (1896); v. Tieghem in Bull. Soc. bot. Fr. 43: 190 (1896); Sprague in F.T.A. 6, 1: 403 (1911); 1034 (1913); in F.C. 5, 2: 127 (1915); Burtt Davy, Fl. Transv. 467 (1932); Balle in F.C.B. 1: 375 (1948); Adamson in Fl. Cape Penins. 343 (1950); Kidd, Wild Flow. Cape Penins. pl. 39, 2 (1950); Letty, Wild Flow. Transv. pl. 61, 1 (1962); Batten & Bokelmann, Wild Flow. E. Cape Prov. pl. 55, 5 (1966); Balle in Mitt. bot. StSamml., Münch. 7: 193 (1968) pro parte, excl. *V. schaeferi*; in F.S.W.A. 22: 12 (1968) pro parte, excl. *V. schaeferi*; Ross, Fl. Natal 153 (1972). Type: Cape, *Thunberg* in Herb. Thunberg 23346 (UPS, holo.; PRE, microfiche!).

V. glaucum Eckl. & Zeyh., S. Afr. J. 375 (1830); Enum. 357 (1837). Type: Cape, Swartkopsrivier, *Ecklon & Zeyher* 2276 (PRE!).

V. thymifolium Presl, Epim. Bot. 251 (1851). Type: Cape, sine loc. exact., *Drège* 7651 (K!).

V. tricostatum E. Mey. ex Harv. in F.C. 2: 580 (1862). Syntypes: Cape, between Verleptpram and Orange River Mouth, *Drège* s.n. (TCD!); Orange River, *Zeyher* 747 (TCD!); Namaland and Hopetown districts, *Wyley* s.n. (TCD!).

V. macowanii Engl. in Bot. Jb. 19: 131 (1894). Type: South West Africa, Ubib, *Gürich* 13 (B†).

V. ziziphi-mucronati Dinter, Deutsch-S.W. Afr. 56 (1909), nom. provis.

V. bosciae-foetidae Dinter in Reprium nov. Spec. Regni veg. 24: 368 (1928), nom. nud.

V. pauciflorum sensu F. Bolus, L. Bolus & Glover in Ann. Bolus Herb. 1: 102 (1915).

Leafy, monoecious shrubs of relatively small size, often forming small rounded clusters less than 0,5 m high, mostly light, pale green, often glaucous; younger branches 6-ribbed but rounded; basal internodes of younger branches mostly 12–20 × 1–1,5 mm. *Leaves* highly variable, ovate-suborbicular to elliptic-oblong, 8–12 × 4–8 mm, apex obtuse-acute, usually minutely apiculate, base rounded to acute, often 3-nerved from base, usually glaucous; petiole sessile-subsessile. *Dichasia* with central flower staminate, lateral pistillate, occasionally all pistillate, bearing a short (c. 1 mm) peduncle. *Berries* ellipsoid, 4–5 mm long, smooth, orange, forming a pedicel 3–4 mm long at maturity; style persistent. *Flowering* in April, June, and August but probably erratic in this regard; $n=14$. Fig. 17.

Parasitic on many, diverse hosts including species of *Acacia, Boscia, Carissa, Ehretia, Euclea, Grewia, Maytenus, Olea, Passerina, Rhamnus, Rhigozum, Rhus, Salix,* and other mistletoes (*Tapinanthus* and *Viscum*), and probably many others. The most prevalent and widespread mistletoe in Southern Africa (including Rhodesia) found under a wide variety of ecological conditions (Map 19).

Vouchers: *Acocks* 2288; *Marloth* 8473; *Galpin* 1820; *Ecklon & Zeyher* 624.

The most widespread and polymorphic species of *Viscum* in Southern Africa. A number of the variations have been given names (see synonymy) but none appear to warrant formal taxonomic recognition. A possible exception could be *V. tricostatum* which was named from populations in the north-west Cape and southern South West Africa. A specimen from this area did exhibit elongated internodes and narrower, longer, and darker leaves, than typical *V. rotundifolium*, but more extensive field studies must be completed to determine the status of these plants.

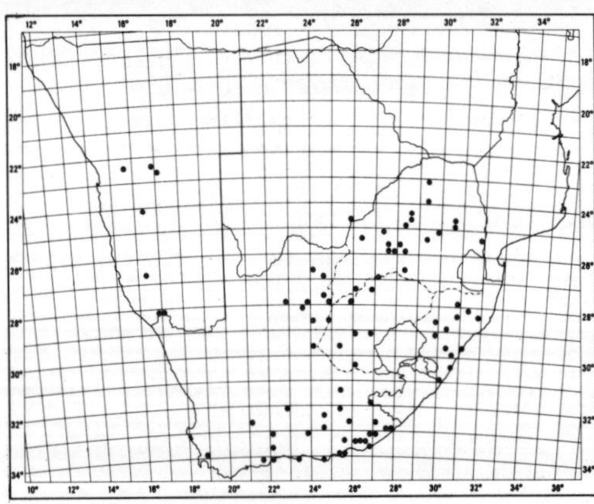

MAP 19.—**Viscum rotundifolium**

12. **Viscum pauciflorum** *L.f.*, Suppl. 426 (1781); Thunb., Fl. Cap. ed. Schultes 154 (1823); DC., Prodr. 4: 285 (1830); Eckl. & Zeyh., Enum. 357 (1837); Harv. in F.C. 2: 579 (1862); Sprague in F.C. 5, 2: 126 (1915). Type: Cape, Karoo beyond Bockland, *Thunberg* in Herb. Thunberg 23344 (UPS, holo.; PRE, microfiche!).

V. eucleae Eckl. & Zeyh., Enum. 357 (1837); Sprague in F.C. 5, 2: 126 (1915). *V. pauciflorum* var. *eucleae* (Eckl. & Zeyh.) Harv. in F.C. 2: 580 (1862). Type: Cape, Winterhoekberge and Onderbokkeveld, *Ecklon & Zeyher* 2275 (SAM!).

Leafy, monoecious shrubs of small to moderate size, perhaps 0,5–0,75 m high; often dark grey and lightly glaucous; younger branches usually 6-ribbed; older branches rounded; basal internodes of younger branches highly variable, (10–) 15–25 (–35) ×1–2 (–3) mm, only slightly (if at all) dilated at the nodes. *Leaves* mostly elliptic-oblong to obovate, 15–25 ×8–12 mm, apex obtuse-rounded, often minutely apiculate, base usually obtuse, often 3-nerved basally, relatively thick and coriaceous; petiole subsessile to 1 mm long. *Dichasia* with central flower typically staminate, lateral pistillate, but sometimes all 3 flowers pistillate, 1–2 per axil, shortly (c. 1 mm) pedunculate. *Berries* mostly ovoid, 4–5 mm long, smooth, orange, forming a pedicel at maturity c. 3 mm long (the peduncle also sometimes expanding to 2–3 mm); style minute, persistent. *Flowering* in June and November, but probably erratic in this regard; $n=14$. Fig. 17.

Parasitic on species of *Euclea, Maytenus,* and *Rhus* in the mountains of the south-western Cape Province (Map 20).

Vouchers: *Boucher* 2003; *Eicker & students* 114; *Wiens* 5411.

This species is closely related to *V. rotundifolium* with which it occurs sympatrically in the mountains north of Worcester.

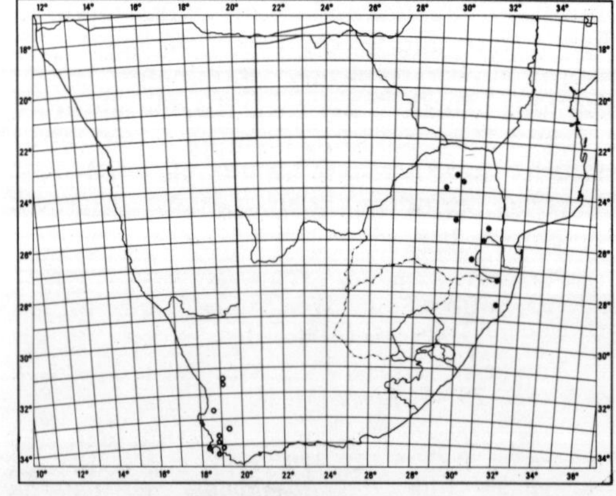

MAP 20.—○**Viscum pauciflorum**
●**V. subserratum**

13. Viscum subserratum *Schltr.* in J. Bot., Lond. 34: 504 (1896); Sprague in F.C. 5, 2: 124 (1915); Burtt Davy, Fl. Transv. 466 (1932); Ross, Fl. Natal 153 (1972). Type: Transvaal, near Barberton, *Galpin* 452 (K!; PRE!).

V. galpinianum Schinz in Vjschr. naturf. Ges. Zürich 49: 179 (1904). Type: same as for *V. subserratum*.

Leafy, dioecious shrubs of moderate size, mostly 1 m or higher, sometimes pendulous; younger branches flattened, usually 6-ribbed; older branches rounded; basal internodes of younger branches rather variable, mostly (15–) 20–30 (–35) × 2–4 mm, nodes often dilated. *Leaves* mostly obovate to broadly oblanceolate, 20–35 × 10–20 mm, apically rounded, cuneate into base, 3 (–5)-nerved from base (sometimes faintly), margin (especially on older leaves) often minutely serrulate; petiole 1–3 mm long. *Staminate flowers* in sessile, axillary dichasia (occasionally 2-flowered), 1–4 per axil. *Pistillate flowers* solitary in bracteal cups, shortly pedunculate (2–3 mm), 1–4 per axil. *Berries* ovoid, 5–6 mm long, warty when young, truncate apically, dull yellow-orange, nearly smooth at maturity and developing a short (1–2 mm), stout pedicel; style persistent, 2–3 times longer than wide, broader at base; stigma capitate. *Flowering* from April–October, possibly sporadically throughout the year; $n=11$. Fig. 17.

Parasitic on species of *Kirkia*, *Maytenus* and *Rhus* in the central and eastern Transvaal south through Swaziland to northern Natal (Map 20).

Vouchers: *Thorncroft* 2227; *Wiens* 5255.

This species is often confused with *V. spragueanum* (see discussion under that species).

14. Viscum crassulae *Eckl. & Zeyh.*, Enum. 357 (1837); Harv. in F.C. 2: 580 (1862); Sprague in F.C. 5, 2: 129 (1915). Type: Cape, near Bothaberg, *Ecklon & Zeyher* 2277 (MEL!; S!).

V. euphorbiae E. Mey. ex Drège, Zwei Pfl. Doc. 61: 229 (1843), nom. nud.

Leafy, dioecious semi-shrubs of rather small size, rarely exceeding 0,2–0,3 m high, often stout with age, highly succulent and dark green when fresh. *Stems* rounded (extremely brittle when dried); basal internodes of younger branches 15–20 × 3–4 mm; nodes usually swollen, especially in older stems. *Leaves* generally obovate-orbicular, mostly 6–12 × 5–8 mm, highly succulent when fresh (wrinkled and distorted when dried and easily broken from the stems); petiole subsessile to 2 mm long, relatively broad. *Staminate flowers* in dichasia, 1–3 per axil. *Pistillate flowers* solitary (occasionally 2) in axils. *Berries* ovoid, 5–6 mm long, sessile, smooth, bright orange, subtending bracteal cup developing a short, stout peduncle c. 2 mm at maturity; style persistent. *Flowering* in July and August; $n=12$. Fig. 16.

Parasitic primarily on *Portulacaria afra*, rarely on succulent *Euphorbia* spp. in the eastern Cape from the Sundays River Valley to the Patensie area.

Vouchers: *Dyer* 1189; *Marloth* 6838; *Wiens* 5374.

A highly distinct and unusual species but rarely collected because of its inconspicuousness on the host; however, it is not uncommon in dense stands of *Portulacaria*, its usual host. With experience it can be detected easily, and at some distance, by virtue of its relatively dark green colour.

15. Viscum obscurum *Thunb.*, Prodr. 31 (1794); Fl. Cap. ed. Schultes 154 (1823); DC., Prodr. 4: 285 (1830); Eckl. & Zeyh., Enum. 357 (1837); Harv. in F.C. 2: 579 (1862); Sprague in F.C. 5, 2: 125 (1915); Batten & Bokelmann, Wild Flow. E. Cape Prov. pl. 55, 4 (1966); Ross, Fl. Natal 153 (1972). Type: Cape, Slangrivier near Clute, *Thunberg* in Herb. Thunberg 23339 (PRE, microfiche!).

V. obscurum var. *brevifolium* Harv. in Fl. Cap. 2: 579 (1862). *V. brevifolium* (Harv.) Engl. in Bot. Jb. 20: 131 (1894). Syntypes: Cape, Swartkops River, Albany, Kafferland, *Ecklon & Zeyher* 2272 (SAM!); *Zeyher* 2700 (PRE!); Kaymansgat, *Drège* s.n.

V. obscurum var. *longiflorum* Harv. in F.C. 2: 579 (1862). Type: Cape, near Swellendam and on Chumiberg, *Ecklon & Zeyher* 2273 (SAM!).

Aspidixia bivalvis v. Tieghem in Bull. Soc. bot. Fr. 43: 192 (1896). *Viscum bivalve* (v. Tieghem) Engl. in Pflanzenfam., Nachtr. 1: 140 (1897). Type: Cape, sine loc. exact., *MacOwan* s.n.

Viscum pauciflorum sensu Drège, Zwei Pfl. Doc. 129 (1843).

Leafy, dioecious shrubs, up to 1 m or higher, sometimes glaucous, mostly pale green; younger branches usually 6-ribbed; older branches rounded with often swollen nodes; basal internodes of younger branches (25–) 35–40 × 1–2 mm, often dilated at nodes. *Leaves* oblanceolate-obovate, highly variable in size (15–) 25–35 (–50) × 8–15 mm, rounded apically, cuneate into base, 3-nerved basally (sometimes only faintly so), occasionally glaucous; petioles rather indistinct when merging into blade, c. 3–4 mm long. *Staminate flowers* in sessile axillary dichasia, 2–3 per axil (sometimes accessory flowers developing laterally and increasing the number to 5). *Pistillate flowers* solitary in bracteal cups, these born singly, or in fascicles of 2–3 per axil. *Berries* ellipsoid-globose, 6–7 mm long, developing a relatively long (4–5 mm) pedicel slightly widening towards the apex at maturity, smooth, whitish cream to light pink; style persistent. *Flowering* from June through July; $n=15$. Fig. 18.

Parasitic on numerous hosts including species of *Acacia, Maytenus, Cussonia, Garcinia, Manilkara, Olea, Populus, Prunus, Rhus, Salix, Schotia* and *Trimeria* and occurring from south central Cape Province north-eastward to central Natal; apparently disjunct in southern Kruger National Park (Map 21).

Vouchers: *Acocks* 8962; 11985; *Marloth* 11951; *Galpin* 2483.

A highly variable species, especially with respect to leaf size. *V. obscurum* var. *longifolium* and *V. obscurum* var. *brevifolium* are based on atypical leaf types which appear sporadically and do not appear to be geographically consistent. Var. *brevifolium* is apparently a form common on exotic species (*Salix*) to which it could probably not have become adapted since their introduction. These leaf variations thus seem likely to reflect variations in the host or the ambient environment and as such should not receive taxonomic recognition.

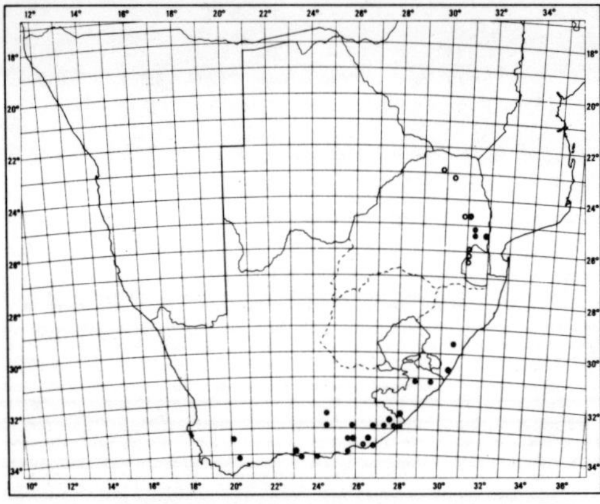

MAP 21.—●Viscum obscurum
　　　　　○V. oreophilum

16. **Viscum oreophilum** *Wiens* in Bothalia 12: 423 (1978). Type: Swaziland, Mbabane, *Compton* 27859 (PRE, holo.!; NBG!).

Leafy, dioecious shrubs of moderate size, mostly 0,5–1 m high, mostly dark green; younger branches flattened, 6-ribbed, those below leaves sometimes more prominent; older branches rounded; basal internodes of younger branches (15–) 25–35 × 3–4

mm; nodes often dilated. *Leaves* broadly oblanceolate-obovate, (25–) 30–40 × 12–20 mm, apically rounded-obtuse, cuneate into base, conspicuously 3-nerved from base; petiole rather indistinct where merging into blade, c. 3–4 mm long. *Staminate flowers* in sessile dichasia, mostly 1 (2) per axil but up to 8 dichasia on terminal nodes. *Pistillate flowers* solitary in bracteal cups, these borne singly in axils of younger stems. *Berries* ellipsoid to globose, 5–6 mm long, developing a relatively long (4 mm) pedicel at maturity, smooth, bright orange; style persistent. *Flowering* in August and September; $n=14$. Fig. 18.

Parasitic on *Pterocelastrus* sp., the only known indigenous host (also on *Prunus persica*). Recorded from the highlands of Swaziland, the adjoining eastern Transvaal and the Soutpansberg (Map 21).

Vouchers: *Wiens* 5260; 5344; 5460.

The species superficially resembles *V. obscurum* and *V. pauciflorum* vegetatively, but is distinct from the former by the shorter, cylindrical pedicel and the bright orange fruit. The dioecious condition easily distinguishes it from *V. pauciflorum* which is monoecious. In *V. oreophilum* the union between host and parasite is often so indistinct that without dissection it is often difficult to distinguish between the stem of host and parasite.

17. **Viscum minimum** *Harv.* in F.C. 2: 581 (1862); Sprague in F.C. 5, 2: 135 (1915); Batten & Bokelmann, Wild Flow. E. Cape Prov. pl. 54,4 (1966). Type: Cape, Albany, *Barber* s.n. (K!).

Aspidixia minima (Harv.) v. Tieghem in Bull. Soc. bot. Fr. 43: 192 (1896).

Leafless, monoecious, minute herbs, mature shoots (including the terminal inflorescence) c. 3 mm high; the apparently single *internode* bearing a terminal and often 2–3 lateral, peduncled dichasia. *Dichasia* with central flowers staminate, lateral pistillate. *Berries* elliptic-rounded, 7–8 mm high, developing a pedicel c. 2 mm long at maturity, smooth, bright orange; crown-like sepals often persistent to maturity; stigma caducous. *Flowering* in June and July, possibly longer; $n=14$. Fig. 18.

Parasitic only on *Euphorbia polygona* Haw. and *E. horrida* Boiss. in the eastern Cape region.

Vouchers: *Marloth* 14131; *Pole Evans* H18268; *Wiens* 5383.

A remarkable species of *Viscum*, which is almost impossible to find when not fruiting. It is comparable in size to another viscaceous parasite, *Arceuthobium minutissimum* Hook. f., which J. D. Hooker (Fl. Br. Ind. 5: 227; 1886) believed to be one of the smallest dicotyledonous plants.

INDEX

* An asterisk signifies taxa not naturalized in Southern Africa; synonyms are in italics.